BEYOND PRESERVATION

BEYOND PRESERVATION

RESTORING AND INVENTING LANDSCAPES

A. DWIGHT BALDWIN, JR.,
JUDITH DE LUCE, AND
CARL PLETSCH,
EDITORS

UNIVERSITY OF MINNESOTA PRESS • MINNEAPOLIS • LONDON

The map of North Carolina on p. 169 is reprinted with permission of Louisiana State University Press from *Historical Geography of the North Carolina Outer Banks* by Gary S. Dunbar. Copyright 1958 by Louisiana State University Press. Copyright 1986 by Gary S. Dunbar.

Published by the University of Minnesota Press
2037 University Avenue Southeast, Minneapolis, MN 55455-3092
Printed in the United States of America on acid-free paper

Printed on recycled paper (50% recycled/10% post-consumer)

Library of Congress Cataloging-in-Publication Data

Beyond preservation : restoring and inventing landscapes / A. Dwight
Baldwin, Jr., Judith de Luce, and Carl Pletsch, editors.
    p. cm.
    Includes bibliographical references and index.
    ISBN 0-8166-2346-5. — ISBN 0-8166-2347-3 (pbk.)
    1. Restoration ecology—Congresses.   2. Landscape protection—
Congresses.   3. Man—Influence on nature—Congresses.   I. Baldwin,
A. Dwight.   II. de Luce, Judith.   III. Pletsch, Carl.
QH541.15.R45B48   1993
333.73'153—dc20   93-4953
                                                           CIP

Contents

Acknowledgments

We are grateful for the help and support of those individuals, institutions, and organizations without whom this book would not have been possible: Miami University, the Sigma Chi/William P. Huffman Scholar-in-Residence Committee, and our colleagues John Klink, Carl Jantzen, Martin Dulgarian, and Ronald Spielbauer, whose labor and support were essential for the continuation of the symposium series "Reconstructing Past Landscapes." The College of Arts and Science and the departments of classics, geology, and history have all played a part in the production of this book. We are particularly indebted to Jeri Schaner for her hard work and infinite patience in preparing the manuscript, and to Shirley Jones, who was instrumental in preparing the original transcripts of the symposium "Beyond Preservation: Restoring and Inventing Landscapes." Finally, the reviewers of the manuscript not only encouraged us to keep working on the book but made valuable suggestions that improved our original conception. Any oddities or infelicities that remain are strictly our own.

A. Dwight Baldwin, Jr.
Judith de Luce
Carl Pletsch

PART I

BEYOND PRESERVATION

Introduction: Ecological Preservation versus Restoration and Invention

A. Dwight Baldwin, Jr., Judith de Luce, and Carl Pletsch

In the past several decades interest in the fate of the natural environment has caused a proliferation of associations, professional journals, popular magazines, and successful books. People around the world have been galvanized by concerns about pollution, environmental degradation, and resource depletion. Scholars in nearly every field have turned their attention to the problems of the environment. In addition, ecology has come of age as an academic discipline attempting to synthesize knowledge from many other, more specialized, areas.

The Ecological Crisis as an Intellectual Crisis

Intellectual positions and programs have also proliferated. From Bill McKibben's widely published announcement of *The End of Nature* (1989), according to which nature has already been irretrievably lost to human contamination, to James Lovelock's "Gaia hypothesis" (1979), according to which the terrestrial biosphere is a living creature capable of compensating for the excesses of its members, myriad new ways of conceptualizing our ecological dilemmas have emerged. These ideas range from suggestions that humans should withdraw from nature entirely, through various models for sustainable agriculture and development, to reclamation and restoration projects for damaged ecosystems, and even to studies of how we might transplant terrestrial nature to other planets. In this welter of proposals, there is general agreement that *something* must

3

be done, but novel proposals do not add up either to a scientific paradigm or to an effective strategy for dealing with our ecological crisis.

The profusion of ecological proposals reveals more than a lack of theoretical agreement. Our faith in scientific and technological progress has been criticized from all quarters. Carolyn Merchant offers a feminist critique in her book *The Death of Nature: Women, Ecology and the Scientific Revolution* (1980), and there are Marxist, anarchist, and other analyses as well. Our old and intrusive paradigm of science and technology as means of manipulating nature for human purposes is being called into question, without our having found a new frame of reference for relating human activities to the rest of planetary life. Apparently this is a transitional moment when the most varied and exciting hypotheses compete for attention, without fully capturing our vision or our commitment.

One reason for our current intellectual predicament is the increasing awareness of the ecological havoc caused by our species. Since the industrial revolution we have witnessed humankind's progressively more rapid alteration of nature. Consequently we are experiencing a general sense of despair at our inability to act creatively to remedy the environmental problems we have created and continually exacerbate.

We are also gradually recognizing the grand effects of human actions in the past. Ecological historians such as Alfred Crosby (1986) have revealed drastic changes in global ecology going back five hundred years to the first European voyages across the oceans and the ensuing colonization movement, adding a dimension of "ecological imperialism" to that migration. An unintended consequence of the European expansion was the transformation of the biota of the Americas, Australia, New Zealand, and other non-European areas in the temperate zone as a result of the bacteria, plants, and animals brought by the European colonists. Other studies, including William Cronon's *Changes in the Land* (1983), have shown that even such paleolithic societies as native American tribes — traditionally thought to have had a harmonious relationship with nature — had profound and often negative influences on the environment.

Our growing acknowledgment of the impact of humans on nature may be producing more than cognitive confusion: a debilitating sense of species guilt threatens to inhibit our ability to deal with environmental problems. If humans have always shaped the natural world to suit themselves, and if we have always been upsetting the ecological balance and are only doing so much more decisively at present, perhaps the end of nature was inevitable, and necessarily our fault. For some, this new version of original sin is seductive; for others, the knowledge of how implicated we are in environmental degradation leads simply to despair.

In this time of heightened ecological awareness and anxiety, the conservationist pieties of the generations who created the national park system in the United States no longer hold. Some writers have even begun to question the ideology of preservation. Their logic is obvious: we can see how current development has transformed so much of nature in our lifetime; we are learning how nature had already been reshaped by our species in the past; and we are realizing that there is not much left to preserve in its pristine state anyway. Furthermore, we are less and less clear about what it would mean to preserve nature.

Preservation still seems a good idea where it is both feasible and meaningful, but it is inadequate as a comprehensive solution to our worldwide ecological problems. In the best of circumstances, preservation is applicable only to the limited portion of the earth that has not already been tampered with—and even those areas are menaced by people and states that sense far more acutely an immediate need to use the land rather than to preserve it. Even traditionally preservationist groups such as The Nature Conservancy have had to extend their mission beyond the once-hallowed goal of wilderness preservation. Reclamation and restoration of damaged lands are now common projects for these types of organizations.

Beyond the practical economic and political difficulties, and the scarcity of wilderness areas available for preservation, a nagging philosophical issue with the idea of preservation persists. The definition of nature includes evolution by natural selection, which obviously entails change. If nature is ever changing, preservation is a strategy for making nature stay the same and is inherently problematic because nature is not static. From this perspective, preservation is not natural, even in those areas still in relatively pristine condition. Nature is a volatile and dynamic system that does not lend itself to preservation as works of art do. Even as we attempt to prevent ourselves from ruining what is left of pristine nature, we doubt whether what we do is appropriate in the long term.

This may seem a sophistical objection to some. After all, if we were to preserve a large enough tract of pristine nature and prevent human intervention, perhaps evolution would proceed unimpeded there. But that raises what is perhaps the most basic philosophical question about preservation, conservation, or any alternative philosophy of ecology: namely, are we humans "members" of nature or not? If we understand ourselves to be other, different from, and opposed to nature, it makes sense to attempt to protect nature, to isolate and preserve portions of nature from human incursions. The theory of preservation in its strictest sense is founded on this premise. The same belief in the opposition of the human and natural underlies McKibben's argument that nature is dead and gone. It is also the basis of the guilt that might prevent us from acting creatively on nature's behalf during this critical time.

If, however, we consider ourselves as one among many creatures in a grand community of species, although perhaps the most privileged, then the prospect is altogether different. Inextricably entangled with the rest of nature, we are products of evolution like everything else, and inevitably affect all the other elements of nature. This is the premise of both Frederick Turner and William Jordan III. Jordan, a botanist by training, is committed to ecological restoration, which maintains that humans are members of nature, and that along with our unavoidable influence upon the community of species we also have a large degree of responsibility for the whole. At this juncture our responsibility is to discover how to restore what we have been destroying. Turner is a poet and essayist fascinated with the actual invention and construction of new landscapes. In his writing he has advocated a more ambitious human effort to shape nature. He does not even shrink from the idea of terraforming Mars, a subject that recently found a place in the science section of the *New York Times* (October 1, 1991) and in *Nature* (August 1991). As different as the projects of landscape restoration and invention are, they share the fundamental notions that human beings are part of nature, not separate from it, and that removing ourselves from nature is not an option.

Jordan and Turner advocate that we go beyond preservation to take an active role in restoring and actually constructing our landscape with creative ecological ends in mind. Their operating assumptions that humankind is part of nature and thus has an inevitable impact upon it, and their injunction that we accept the responsibility to use our science to make this an ecologically sound and beneficial influence, set them apart from most of today's writers on the environment.

Many environmentalists bridle at the proposals of Turner and Jordan. To preservationists in particular, their writings appear threatening: Jordan and Turner endanger our ability to save nature by suggesting alternatives to preservation. Yet the writings of Turner and Jordan appeal to a broad constituency. They touch on a variety of issues, from the philosophical to the practical, and are neither mystically optimistic like James

Lovelock nor despairingly righteous in the manner of Bill McKibben. They do not advocate turning back the clock to an earlier way of life, as does Wendell Berry (1977, 1987, 1990). Rather, Jordan and Turner urge us to make maximum use of everything we can know about nature to ensure the continuation and evolution of life on earth, and perhaps even to propagate it elsewhere. Their initiatives either inspire or provoke the widest assortment of people working on and concerned with the ecological crisis.

"Restoring and Inventing Landscapes"

Jordan and Turner offer their ideas as a test of how one basic paradigm might survive the rigorous criticism of scholars from a broad range of disciplines. In that spirit they were invited to Miami University in 1990 for a multidisciplinary symposium on landscape restoration and construction.

"Restoring and Inventing Landscapes: Beyond Preservation" was the eighth symposium in a series begun in 1984 on the theme of reconstructing past landscapes. These symposia all involved "reconstruction" in the intellectual sense of recovering the properties of a variety of landscapes through archaeological, geological, historical, or literary research and analysis. They confronted directly the physical remains of the past and sought to reconstruct that past by understanding or interpreting the vestiges of a landscape no longer visible.

The 1990 meeting addressed the literal reconstruction of a landscape using prairie restoration as a case study. Thus, in the terminology developed by Meinig (1979) to identify various types of landscapes, the organizers of this symposium elected to view landscape as an artifact, one so altered by human activity that "pristine" nature had been eliminated. The symposium considered the concept of landscape restoration in this context of landscape as an artifact.

The symposium confronted a number of contemporary issues. Can we reconstruct the past by trying to restore our landscapes? Is restoration biologically feasible? What are the costs of landscape reconstruction, and is this a legitimate goal of society? What are the ethical, political, and social consequences of this kind of work, and what are the competing claims of restoration and preservation? Finally, a field trip to a restored prairie near Dayton, Ohio, highlighted the techniques and consequences of such landscape reconstruction.

To provide a focus for the symposium, William R. Jordan III and Frederick Turner gave keynote addresses. They come to landscape restoration from very different disciplinary backgrounds, and the disparity of their training and their professional careers indicates how interdisciplinary restoration is.

William R. Jordan III was trained as a botanist and has become a writer and editor in the cause of ecological restoration. He is currently on the staff of the University of Wisconsin-Madison Arboretum, the location of an extensive tract of restored tall-grass prairie. The arboretum's prairie is perhaps the longest-lasting experiment in ecological restoration in the United States, and the arboretum has become an important center of restoration research. Jordan coedited a substantial collection of scientific papers, *Restoration Ecology: A Synthetic Approach to Ecological Research* (1987), and has been editor of the journal *Restoration & Management Notes* since it was founded in 1981. He is a leading figure in both the theory and practice of ecological restoration.

Frederick Turner, Founders Professor of Arts and Humanities at the University of Texas at Dallas, is a poet, literary scholar, and essayist. His essays in *Harper's Magazine* (1985, 1988) have proven to be among the most persuasive arguments for ecological restoration; in fact, these essays are the best existing justification for an active human role in shaping nature. Amid Turner's extensive body of published poetry is a grand narrative, *Genesis, an Epic Poem* (1988), in which an "ecotheist" terrestrial regime, governed by a priestly elite dedicated to a theology of preservation, contests the right of a rebel band of humans to terraform the planet Mars. This poem is a visionary attempt to discern the future of the human relationship to nature. In *Genesis*, Turner dramatizes the disparity between preservation on the one hand and restoration and invention on the other.

Consistent with their earlier writings, the presentations of Jordan and Turner at the 1990 symposium challenged the idea of preserving nature. They interpreted the ambition to preserve the pristine remnants of nature as a sterile and indeed impossible approach that denies the status of "nature" to the vast majority of the earth's surface, because it has already been altered or contaminated, and requires humans to absent themselves entirely from uncontaminated areas. Jordan and Turner called instead for recognition of humankind as one more natural species, and one that has always intervened in the rest of nature. They urged that we use our intelligence and our creative faculties to manage nature by restoring damaged landscapes and creating mutually beneficial relationships among all species.

•Landscape reconstruction involves its practitioners in actively shaping (or reshaping) the natural world, creating (or re-creating) communities of species that can live together in an ongoing, self-sustaining way. Turner and Jordan advocate not only researching an ecosystem, but cultivating, manipulating, and even using it. Prairie restoration is the best-known example, and at the symposium Jordan put it forward as exemplary. In Wisconsin in the 1940s, Aldo Leopold and his colleagues searched the roadsides and railroad rights-of-way throughout the Midwest for native grasses that had escaped the plows and cultivators of farms. They gathered what they found and attempted to re-create the prairie on exhausted and eroding farmland. They had some success at first, but it was not until they realized that fire was an important ingredient of the prairie and then started to burn periodically that they got something approaching the original. This hands-on project involved much trial and error but yielded remarkable results, not only in the form of a gorgeous restored prairie that is now part of the University of Wisconsin-Madison Arboretum, but also in scientific understanding of prairies and of the restoration process.

By now prairie restoration is so successful that prairie projects are springing up across the Midwest. State highway departments, including the Ohio Department of Transportation, are experimenting with prairie grasses along the interstate system. There are well-publicized efforts at restoring savannas, wetlands, rain forests, and other ecological systems as well. Jordan and the Society for Ecological Restoration are pioneering in all these areas, and the practical future of this approach to the environment seems grand.

Turner's idea, expressed in *Genesis*, of creating an earthlike landscape on Mars is grandiose and far from implementation, but it is not just a poet's fantasy. Many scientists are skeptical, but some, including several at NASA, are enthusiastic. The obstacles are immense and the time required to create a more earthlike atmosphere on Mars may be tremendous. But given what modern science has achieved already, it seems foolish to say this is impossible. Perhaps more germane than musing on the limits of science, we should notice that, like restoring earthly landscapes, terraforming another planet is a logical extension of the human habit of gardening.

Whether terraforming Mars is desirable and whether it is a good way to spend our resources are important questions. Some argue that the idea of restoring earthly landscapes and creating martian ones will only give succor to those who want to continue abusing the terrestrial environment. Pollution will be harder to police if it becomes common to think that we can easily restore what gets eroded and degraded, or escape to Mars if we manage to corrupt this planet entirely. But whatever the pitfalls, and whatever our reservations about landscape restoration and construction, these ideas clearly go beyond preserving what is left of nature. They hold out the prospect of actually doing something about the ecological crisis, and that may be their most powerful appeal.

The vehement rejection of the philosophy of preservationism by Jordan and Turner provoked much resistance from the audiences and panels of faculty who discussed their presentations at the symposium. Most acknowledged the necessity of restoring as well as preserving nature, but few were willing to accord restoration a higher importance than preservation or to reject preservation altogether. In retrospect it seems that most of the audience and many of the contributors to this volume were intrigued by the practical project of restoration, whether of tall-grass prairies or tropical forests. They saw preservation and restoration as compatible projects that could be pursued simultaneously in different locations. The general resistance to restoration focused on the theoretical pretensions of Turner and Jordan. Few symposium participants saw the need to pit preservation against restoration theoretically, to accept that they are mutually exclusive, or to choose between them. Many of those who would choose preferred preservation.

Given this background to the responses at the symposium, readers of this volume may want to consider, first, if ecological restoration is not a practical necessity. Second, we may wonder about the compatibility of restoration and preservation. Third, if they are compatible in practice, can the theories be reconciled? Finally, whether or not we are prepared to choose between the theories and presuppositions of preservation and restoration, we may ask which is the more viable ideology for the future, and what our other alternatives might be.

As Jordan and Turner depict constructing and reconstructing ecosystems, these projects tax the human imagination and aspiration, demand refined techniques and hard work, and provide opportunities to measure progress—all unlike preservation. Unless we are to renounce further progress in science and technology, and reverse secular trends in economic development, population growth, progress toward more equal standards of living, and so on, it seems that the techniques of landscape construction and reconstruction will become more rather than less important to us. The pressure of events is far more likely to force us to do something than to motivate us collectively to refrain from what we have done for centuries. In fact, if we are now coming to realize that our species has always interfered with nature, then involving ourselves with nature in this new way that we term restoration may be more consistent with the character of our species than preservation. To restrain our interference with nature may be the one impossible option. Yet we lack a shared paradigm for our relationship to the earth and the rest of life, a template to guide our intervention.

Multidisciplinary Reflections

A combination of ideas found in the writings of Jordan and Turner—based on the membership of humankind in nature, and demanding an active and indeed intrusive effort on our part to restore, reconstruct, and even invent new landscapes—is one candidate for a new ecological paradigm. For that reason this volume presents essays of Jordan and Turner together with comments and critiques from a variety of scholars at Miami University. The proximity of these scholars has encouraged ongoing dialogue about the environment, preservation, and restoration. We hope that the compilation of these essays will provide the impetus for broadly based discussions of these issues as they arise in other parts of our country and the world.

These critiques bring the perspectives of many disciplines to the topic of restoration—very appropriately, as landscape itself is not a single phenomenon and cannot be studied adequately from only one approach. Individual chapters in this book work together to integrate various disciplinary methods into a reflection on a common theme. Taken together, the chapters, like Meinig's combined views of landscape, present complex views of landscape restoration. The essays have been divided into three broad categories: theoretical issues of restoration, case studies of the application of restoration/reclamation/preservation theory and techniques, and reflections on the implications and consequences of environmental restoration.

The section on theoretical issues begins with a chapter by G. Stanley Kane that addresses the ethics of restoration. He raises two questions entailed in the controversy between restorationists and their critics: Is restoration of nature desirable? Is it possible? He then identifies the philosophical issues involved in restoration, indicates alternative ways to think about them, and shows how different positions lead to different conclusions about restoration. Although he regrets that restoration offers no protection for the environment against continuing social and ecological disasters caused by humans, he admits that attempts at restoration are an inescapable obligation. Kane concludes, however, that the disagreements between restorationists and preservationists are so fundamental as to be irreconcilable. In his view preservation is the higher priority.

Intellectual historian Carl Pletsch treats the relationship of humans to nature in terms of sovereignty: Do we humans have rightful dominion over nature? If we assert our sovereignty, how will we replace the "law of nature" from which we have derived virtually all of our norms since the seventeenth century? Paradoxically, although with the coming of the environmental movement we have begun to recognize the importance of restoring damaged landscapes, humans remain ill equipped theoretically and morally to deal with this responsibility.

Gene E. Willeke provides an environmental scientist's perspective on the need to develop some societal consensus of the goals and purposes of restoration. He begins with the observation that in the presence of Turner and Jordan, one is reminded that the biologist, when contemplating restoration and invention, may become imbued with the poetic to the point that there is too much deviation from natural principles, and that the poet may come to believe that the biologist understands more than the discipline actually allows.

Dora G. Lodwick concludes this section by examining the changing worldviews and subsequent ideologies about nature since the 1960s. The environmental issues that the well-educated upper class might determine as important do not correspond to those identified by the urban poor. She argues that whoever provides future leadership for addressing the public's concerns about the environment will need to recognize that those concerns vary according to race, ethnicity, gender, and class.

The next section of essays presents a series of case studies from contexts quite different from the prairie restoration highlighted by Jordan. Gary W. Barrett and Orie L. Loucks argue that biological science is not presently capable of restoring a landscape to its pristine form, and even if it were, the resulting ecosystem would have to be managed like a garden in order to preserve it, a much less efficient process than preserving the natural ecosystems that still survive. Barrett discusses the lessons we need to learn before restoration ecology can evolve into an integrative paradigm that joins theory with practical application for the twenty-first century. Above all, we as a nation cannot restore large-scale systems without simultaneously addressing simultaneously societal problems such as world food production, global climatic change, and human population growth.

Loucks asserts that the art seen in restored landscapes is the art of imitation and that restoration is a process of copying. He uses the Ding-hu Shan Biosphere Reserve in southern China as an example of a "great master" in the form of a natural ecosystem, and then considers the impact of human interference and notes the difficulty of restoring landscapes given our lack of knowledge concerning the interactions of an ecosystem's components. Loucks emphasizes that isolated originals are critical, for they provide the framework for understanding how natural systems function.

Botanist David L. Gorchov is skeptical of the possibility of restoring tropical rain forests where they have been destroyed, but suggests that managing tropical forests for a sustained yield of timber might be a way of reconciling development and conservation. This restoration may not be as easy as first envisioned, however, because seedlings of economically valuable tree species cannot establish themselves as readily as was originally believed. Ultimately he argues that if the goal is to preserve essential species that define the habitat, then preservation must be the foundation of our ecological strategy.

Plant geographer Kimberly E. Medley presents a site-specific example in Kenya of how a forest-restoration project in the Tana River National Primate Reserve would help protect two endangered primates. She argues that, although such forest restoration would in no manner restore the ecological balance that existed prior to the disturbance caused by human activities, selective restoration of food-bearing trees would provide a sustainable resource base for these two species of endangered monkeys. Interestingly, in light of Jordan's and Turner's attention to ritual in local restoration projects, she notes that an area's human residents are essential to the conservation of the valuable botanical and zoological landscape.

John E. Wierwille traces the settlement and exploitation of the Outer Banks of North Carolina as a prime example of how human beings have unwittingly and unwillingly interfered with the natural functioning of a fragile ecosystem. He notes that the disruption of this system threatens the economic well-being of the area. Unfortunately, public awareness generally occurs only after ecological dysfunction is irreversible. He closes with insights into how this ecosystem could be restored and become self-sustaining.

Finally, geologist A. Dwight Baldwin, Jr., suggests that reclamation of land mined for coal, even if not a perfect technique, may be profitable in areas that have been flagrantly abused by strip-mining. He poses a series of questions, including who decides what represents "higher and better use" for mined land; on what basis this decision is made; and what restoration techniques could be employed to create more aesthetically pleasing and ecologically diverse ecosystems. Most mined-land restoration today lacks vision as to what should be the long-term goals of mined-land reclamation.

Restoration and preservation must occur within a human context, so no serious discussion can ignore the implications and consequences of these activities. The choices humans make in their relations with nature not only impact nature, they inevitably reflect the character of particular human societies. Printmaker Ellen Price uses her work on garden statuary to consider our uneasy relationship with nature. In the examples of lawn ornaments she finds an analogy between the restoration of a landscape such as a prairie and of animal life that is "restored" in the manageable, sanitized forms of concrete beasts. These "creatures" reflect our desire for some connection with nature, but they represent a restored nature that mirrors none of our failures to respect and acknowledge the natural systems.

Judith de Luce locates Turner's epic poem *Genesis* within the traditions of cosmogonic mythology and rituals of different cultures. She suggests that because this poem derives and departs from these traditions, it can inspire the reader to think about the motives and implications of human interference in the natural, nonhuman landscape. Humans have arrogated the role of cosmic creator; if we succeed in terraforming a Mars, we will need myths as well as rituals to remind ourselves why we did it, how we did it, and at what cost.

In a related chapter, architect Ann Cline focuses on ritual as she considers landscape not only as a place of actual habitation, but as the site of ritual and performance (in this case the tea ceremony), which provide meaning and structure to human experience. The prairie and the teahouse both may be most useful in the issues they raise about time, pleasure, consumption, and change. Once the prairie landscape has been reconstructed, its next use could be in its ritual inhabitation and the mythology of that act.

Constance Pierce analyzes the unspoken ideology implied by the proposals of Turner and Jordan. She considers the audience or consumer of the landscapes they describe and asks who determines, and how, what is worthy of restoration, and what the cultural effects of those decisions are. She concludes by asking questions that have been suggested in previous chapters. Can rituals be made meaningful? Whose well-being do we compromise in order to restore a prairie, for example, and for whom do we do it?

Finally, historian Jack Temple Kirby begins with the observation that restoration would employ the bulldozers and chemicals of environmental degradation to restore the "pristine" and the "classical" landscape that those machines and chemicals had compromised. But what are the costs? He notes that thus far restorationists "garden" mostly in the mind and remain silent about political economy. This chapter asks what the economic and social consequences of restoration are and how we should deal with them. What of our class system, for example, in which the poor live in greatest danger of the toxicity our culture produces? Kirby concludes by stating that while restoration ecology certainly raises consciousness about our culture's contempt for nature, intellectuals have a greater responsibility than mere awareness.

These essays are all written in response to the positions advanced by Jordan and Turner in favor of ecological restoration and invention. They represent a broad spectrum of disciplines, but responses from still other disciplines might be equally relevant. Some of these essays are positive and some critical, but regardless of the degree of praise or criticism they incorporate many different approaches to restoration and invention. As diverse as these approaches are, they are no more exhaustive than the list of disciplines represented. Together they constitute the beginnings of a conversation on these ideas. In that spirit of conversation we invited Turner and Jordan to respond to the essays, in contributions presented at the end of the volume. We hope that their comments and rejoinders advance the conversation further. Finally, we have concluded the book with a brief review of what we think has been clarified and what remains to be discussed. May this conversation continue!

REFERENCES

Berry, Wendell. 1977. *The Unsettling of America: Culture and Agriculture.* San Francisco: Sierra Club Books.

_____. 1987. *Home Economics*. San Francisco: North Point Press.

_____. 1990. *What Are People For?* San Francisco: North Point Press.

_____, Bruce Colman, and Wes Jackson. 1984. *Meeting the Expectations of the Land: Essays in Sustainable Agriculture and Stewardship*. San Francisco: North Point Press.

Cronon, William. 1983. *Changes in the Land: Indians, Colonists, and the Ecology of New England*. New York: Hill & Wang.

Crosby, Alfred W. 1986. *Ecological Imperialism: The Biological Expansion of Europe, 900-1900*. New York: Cambridge University Press.

Gomez-Pompa, Arturo, and Andrea Kaus. 1992. "Taming the Wilderness Myth." *BioScience* 42: 271-79.

Jackson, Wes. 1980. *New Roots for Agriculture*. San Francisco: Friends of the Earth/The Land Institute.

Jordan, William R., III, Michael E. Gilpin, and John D. Aber, eds. 1987. *Restoration Ecology: A Synthetic Approach to Ecological Research*. Cambridge: Cambridge University Press.

Lovelock, James. 1979. *Gaia, a new Look at Life on Earth*. New York: Oxford University Press.

McKibben, Bill. 1989. *The End of Nature*. New York: Random House.

Meinig, Donald William. 1979. *The Interpretation of Ordinary Landscapes: Geographical Essays*. New York: Oxford University Press.

Merchant, Carolyn. 1980. *The Death of Nature: Women, Ecology, and the Scientific Revolution*. San Francisco: Harper and Row.

Turner, Frederick. 1985. "Cultivating the American Garden: Toward a Secular View of Nature." *Harper's Magazine* 271 (August): 45-52.

_____. 1988. "A Field Guide to the Synthetic Landscape: Toward a New Environmental Ethic." *Harper's Magazine* 276 (April): 49-55.

_____. 1988. *Genesis, an Epic Poem*. Dallas: Saybrook Publishing Company.

"Sunflower Forest":
Ecological Restoration as the Basis for a New Environmental Paradigm

William R. Jordan III

I first encountered the writing of Fred Turner, my partner in this dialogue, in the summer of 1985, when I read his essay "Cultivating the
American Garden" in the August issue of *Harper's Magazine*. The essay
made a profound impression on me, and since then Fred's thinking has
contributed immeasurably to my own work at the University of Wisconsin-Madison Arboretum, and in particular to my thinking about the process of ecological restoration and its implications for the environment
and for our relationship with nature.

Briefly, what Fred was suggesting was that the act of gardening offers
a model for a healthy relationship between human beings and the rest of
nature. His argument, in part, was that the gardener handles nature with
respect but without self-abnegation—that is, he or she manipulates nature intelligently and creatively, benefiting and nurturing plants (and of
course animals as well; we are speaking of "gardening" in a broad and
even metaphoric sense here), while at the same time exercising a wide
range of human aptitudes and leaving a distinctively human mark on the
landscape. This struck me immediately, both because Fred's idea was in
accord with my own experience as an amateur gardener and beekeeper
(activities that I had long felt provide a basis for communion with other
species) and also because it was close to my own thinking about ecological restoration and its implications for the environment and our relationship with it. By the time I read Fred's essay, I had already identified restoration as a form of gardening, and had begun to think that it
represented a model for a healthy relationship between human beings and

the natural landscape. The weakness of Fred's conception, from an environmentalist's point of view, was that it placed little emphasis on that natural landscape, but this was where my own idea of ecological restoration came in. If gardening provides a model for a healthy relationship with nature, then restoration is that form of gardening concerned specifically with the gardening, maintenance, and reconstitution of wild nature, and is the key to a healthy relationship with it.

Shortly after reading Fred's essay I called him and we began a conversation that has continued, on and off, ever since. One result has been the linking of our two lines of thought into the more comprehensive idea that the more general process of *ecosystem construction* provides the basis for healthy interaction between human beings and the rest of nature. The key idea here is that we can best come to understand ecosystems, and to enter into a relationship with them that engages the full array of human activities, by attempting to reconstruct them. Like all forms of agriculture, however, the process of ecosystem construction has two poles: a creative pole, most clearly represented by traditional forms of agriculture, which not only construct ecosystems but create or invent new ones, and a conservative pole, exemplified by the form of gardening we term ecological restoration, the attempt to create ecosystems that resemble as closely as possible natural or historic models.

My purpose in this chapter is to explore the left limb of this axis, and to make a case for the idea that ecological restoration provides a basis—actually, a paradigm—for a healthy, mutually beneficial relationship between ourselves and the natural landscape. I will begin by stating what I consider to be at least some of the essential elements of such a relationship.

First, in order to have a relationship with anything we need the thing itself—in this case the natural or historic ecosystems, the forests, prairies, wetlands, lakes, rivers, dunelands, reefs, and so forth, and all the plants, animals, and abiotic elements, all of which comprise the natural landscape.

Second, we need an ecological relationship with these systems. By this I mean an economic transaction that entails a genuine exchange of goods and services between ourselves and the natural community. This must be reciprocal, or, as Aldo Leopold and others have said, mutually beneficial, involving both taking and giving back.

Third, this relationship must engage all our abilities—those that are innate or "hard-wired" into us by evolution and those that have emerged in the course of cultural evolution. These include our physical, mental, emotional, and spiritual capacities.

Fourth, because one of these abilities is a sense of history, and of history as a kind of progress, or at least change, the relationship must acknowledge and deal with the past — the history of our interaction with a particular landscape, and the deeper history of the general relationship of our species with the rest of nature.

Fifth, because our relations with nature continue to change as a result of ongoing intellectual advances and cultural evolution, the paradigm defining that relationship must also be flexible and capable of a creative expansion and development.

Sixth, we are a language-using, social, and highly self-conscious species, so we need a way not only to explore and redefine the terms of our relationship with nature, but also to articulate and celebrate that relationship in a personally and socially satisfying way.

Now let us turn to the question of how ecological restoration provides a paradigm that satisfies each of these criteria.

1. *The object*. Restoration is difficult and uncertain at best, and the craft of restoration is in its infancy. Even the highest-quality examples, such as parts of Greene Prairie at the University of Wisconsin-Madison Arboretum, or Ray Schulenberg's tall-grass prairie at the Morton Arboretum near Chicago, are defective — that is, they are not precise replicas of their natural counterparts. In fact, some environmentalists insist that restoration is impossible and argue that conservation of natural systems depends, ultimately, on preserving those that already exist. Yet if restoration in the strictest sense is impossible, so is preservation. It is impossible either to stop a living ecosystem from changing or to prevent its change from reflecting our influence. Restoration, however, holds out at least the possibility of conserving the system, not by stopping change, but by directing it, and not by ignoring human influences, but by acknowledging and seeking to compensate for them.

In this sense, then, preservation is impossible and restoration merely more or less difficult. All systems are constantly changing, and as this change reflects at least some degree of human influence, all systems must be supposed to be moving continually toward some novel condition. This effect is especially dramatic here in the Midwest on our tall-grass prairies and oak openings, where the entire native ecosystem has been virtually eliminated as a direct or indirect result of new kinds of human activities. This situation is actually paradigmatic, however, and is true in the final analysis of all ecosystems everywhere—not because we are a peculiar or pernicious species, but simply because, as John Muir said, everything is hitched together so that everything interacts with everything else. Acknowledging our membership in the land communities is the first crucial step toward our reenfranchisement in it.

The consequence is that in the long run the best natural areas—the ones most closely resembling their historic counterparts—will not be those that have simply been protected from human influences (complete protection is impossible) but those that have been in some measure restored through a process that recognizes human influences and then effectively compensates for them. This is already evident for the midwestern prairies and oak openings, and sooner or later will be true of all ecosystems.

This being the case, it is encouraging to keep in mind that this does not necessarily imply a gradual decline in the quality of these systems through the process of copying and recopying. The criticism that restoration is impossible generally applies only in the strictest sense. One cannot duplicate a natural system root hair for root hair and bird for bird, but there is no reason to try to do this. What is called for, rather, is the reassembly of a system that *acts* like the original. This implies not only complete species lists and the reproduction of crucial aspects of community structure, but also the reproduction of function and of dynamics—both ecologically and in the evolutionary sense. In other words, it means not just setting the system up, like a diorama, but actually setting it in motion. It also means, however, setting certain limits to this motion: the system conserved in this way may be supposed to be moving around in a defined zone of change judged to be appropriate for the system. This does imply a certain conservatism (we are concerned, after all, with the conservation of the historic system), which implies continual monitoring and specific measures to, as it were, nudge the system back toward its historic condition. But this approach by no means implies a static conception of the system, or of our relationship with it. In fact, quite the contrary: this approach involves a kind of dynamic equilibrium (within certain, often rather wide, limits) and a perpetual effort to sustain the system against the pressure of change in response to new influences. In the long run, this will be the only way to ensure the existence of classic (and in a sense obsolete) ecosystems in the landscape of the future. This is real conservation, something we will want to do, not everywhere, but in some places. The result, of course, as we invent novel ecosystems, in certain cases including new, genetically engineered species, will increase rather than decrease biological diversity and richness.

   2. The ecological dimension. The real challenge of environmentalism is not to preserve nature by protecting it from human beings or rescuing it from their influence, but to provide the basis for a healthy relationship between nature and culture. What this means most obviously is a working relationship with the natural landscape in which a human individual or community can achieve full citizenship in the biotic community. This is what Leopold had in mind when he called for a "mutually beneficial relationship" between nature and culture, but exactly what such a relationship would actually look like has remained unclear.[1] Presumably it would include an actual ecological interaction with the natural landscape that benefits both it and us—and would do so without requiring us to repudiate the achievements or abandon the accoutrements of civilization. From the point of view of modern environmentalism, however, with its strong sense of distance between humans and nature and its idealization of wilderness as nature "untrammeled by man,"[2] such a relationship has proved inaccessible. What environmentalism has offered instead is a severely limited relationship characterized by an ethic of "minimal impact" and the admonition to "take nothing but pictures; leave nothing but footprints." The concern here is almost exclusively for the landscape and hardly at all for the human participant, and the resulting relationship, though valuable as far as it goes, is extremely attenuated. It is largely nonparticipatory, and engages only a small fraction of human interests and skills. The person is confined to the role of visitor—an observer of nature rather than an active element of the land community. Ironically, such a perspective turns us all—hiker, birder, and strip miner alike—not into members of the community but into users and consumers of the natural landscape.

   This may be useful as *part* of a healthy relationship with the natural landscape, but it falls far short of what we have to accomplish if we are to save the classic ecosystems and share the landscape with them. A comment in Thoreau's *Journal* illustrates the point. Thoreau was deeply concerned with achieving an intimate relationship with nature, and most of his writing is the account of his attempt to do so. Not infrequently he imagines himself literally rejoining the natural community by taking the part of one of its members. In an early entry in his *Journal* he wrote: "Would it not be a luxury to stand up to one's chin in some retired swamp for a whole summer's day, scenting the sweet-fern and bilberry blows, and lulled by the minstrelsy of gnats and mosquitoes? . . . Say twelve hours of discourse with the leopard frog."[3]

Here Thoreau is seeing himself as a turtle or muskrat. The problem is that he doesn't push his own figure far enough—this is not what muskrats or turtles do in a marsh. They don't sit there, watching the sun go overhead. They go about their business, which is the construction and maintenance of the marsh. This, of course, is precisely what the restorationist does. He or she is not merely an observer of the marsh or prairie, but, like the muskrat, a maker of the marsh, a direct participant in its ecology, carrying out business there in the fullest—in fact, in the Thoreauvian—sense of that word, exercising skill and ingenuity, exchanging goods and services, influencing and helping to shape the community, communicating with nature in nature's terms.

Thus the restorationist resolves a dilemma that has troubled—and weakened—environmentalism since Thoreau's time. Through the constructive process of restoration he or she breaks out of the essentially negative relationship with the natural landscape implicit in the preservationist program and establishes a relationship with that landscape that is both positive and mutually beneficial—and does so, moreover, without leaving civilization behind. This leads to a way of solving the practical problem of overuse of natural areas. The traditional approach to this problem is to discourage use and place restrictions on activities, a policy based on the presumption that "use" is destructive or consumptive and necessarily compromises the natural landscape. From this point of view the visitor is just that: a visitor and a consumer at best, and at worst an out-and-out destroyer. The more such visitors there are in a natural area, the more "pressure" will be placed on it, and the more it will decline in quality.

Limiting use is one way to address this problem, but it is only a stopgap measure that does nothing either to satisfy the human hunger for immersion in nature or to deal with the unavoidable problem of ecosystem drift in response to human influence, however subtle. The real key to conservation is not restricting human participation, which is merely another way of fighting nature, but to find a constructive way of participating. Much better than proscribing involvement, then, is to change its sign, so to speak, from positive to negative. The visitor then becomes a positive and contributing rather than a negative, consuming force in the landscape. The range of experiences available in the landscape increases dramatically, and the situation shifts from having too many people using up nature to not having enough to keep it in shape.

I mean this quite literally. For years we at the University of Wisconsin Arboretum believed that our greatest problem was overuse. Too many people were, in that ugly and desperate phrase, "loving us to death." Today we have the opposite attitude: we don't have enough people to keep up with the restoration we could be doing. This is also true in the suburbs of Chicago, where the recovery of prairies and oak openings in the splendid system of preserves surrounding the city has depended almost entirely on a growing cadre of volunteer restorationists.[4] Both of these examples are from heavily used areas, but the principle applies everywhere. Eventually it will be applied to our national parks and other wilderness areas, and it will be their salvation. In my view this restoration will become the principal outdoor activity of the next century, and the result will be the conversion of nature — in its classic forms — from an "environment" into a habitat for human beings.

Briefly, then, restoration is the key to the reinhabitation of nature and, in the long run, to its preservation. In its absence our influence on nature is necessarily consumptive; in the context of a restoration program, however, use becomes the first step in achieving a reciprocal relationship, which is completed in the act of restoration. Restoration in this sense is nothing but the acknowledgment of human influence on the landscape and the attempt to compensate for it in a precise way so that the classic landscape may be maintained.

3. The gamut of human abilities. In his book *The Invisible Pyramid*, Loren Eiseley wrote that human beings must not only reenter the "sunflower forest" of original nature, but they must do so without abandoning the lessons learned "on the pathway to the moon."[5] Eiseley's assertion underscores a crucial weakness of the traditional environmental response to the challenge of reinhabitation: its failure to deal with the full range of human abilities, interests, and values, including those that are the achievements of culture. It is relatively easy to imagine reentering nature destructively on the one hand, or by shedding the accoutrements of civilization, on the other, or by simply leaving behind most of what makes us who we are when we step into the forest. But when we do this we limit our relationship with nature; we cease to be fully ourselves, and this makes nature not our habitat but some "other place" — not a whole world in which we "go and come with a strange liberty in Nature, part of herself," as Thoreau wrote,[6] but just another facility with a specialized purpose, like a bank or car wash.

Restoration meets this problem head-on. As a comprehensive process, it includes traditional nature-oriented activities such as hiking, birding, and botanizing, but also a wide range of other, more participatory activities, including hunting, fishing, gathering, and cultivating. All of these are integrated into an event that is constructive rather than consumptive—as each of these particular activities is in its traditional form. Restoration engages a range of physical, intellectual, social, and emotional faculties and actually entails a kind of recapitulation of cultural evolution, a redeployment of all the skills exercised and achieved by human beings, in Eiseley's phrase, "on the pathway to the moon."

Of special interest is the observation that restoration challenges our understanding of the ecosystem being restored, and so is an effective research technique, a way of raising questions and testing ideas about the systems under construction and (not incidentally) about our relationship with them. This notion is embedded in ecological thinking and practice, and has recently been explored and given the name "restoration ecology."[7] This recognition of restoration as a form of dialogue with nature has important implications. First, restoration then does for ecology what the indeterminacy principle did for physics: it recognizes the researcher as an active participant, interacting with and influencing the system being studied. It also places us in a position to develop restoration as a powerful tool for exploring the ecological aspects of our interaction with nature because, although we can change nature without knowing what we are doing, it is virtually impossible to change it *back* without comprehending in some detail both the system and the precise ways in which we have influenced it. Thus restoration brings to our attention aspects of our relationship with nature that otherwise we might not recognize.

4. *The past.* Civilization is characterized by the sense of history and the discovery of cultural change. Archaic peoples, according to Mircea Eliade, had a past that was largely mythic, and they devoted considerable energy to world renewal rites and other activities that had the effect of obliterating history.[8] We, however, know something of history and realize that our relationships with particular landscapes and with nature generally have undergone dramatic changes, especially during the past few thousand years.

Because this awareness is integral to our worldview, presumably it is also an important component of our relationship with nature, and our paradigm must accommodate this. We need the modern equivalent of the world renewal rituals of archaic peoples, not merely to renew the earth in a literal sense (which, in fact, restoration does, offering a fascinating parallel to these classic rituals), but also to explore the past and have access to the experiences of nature that have shaped us as a species, as a culture, as a community, and as individuals.

Ecological restoration offers this opportunity in various ways—or, perhaps more accurately, it provides access to several octaves[9] of historic experience: the immediate experience of the individual in a particular place; the usually longer history of the community and of a particular society or civilization; the still deeper history of cultural evolution; and ultimately the "history" of nature as chronicled by students of evolution and biogeography.

In the first instance, the restorationist revisits history while trying to reverse it. Restoration is, in fact, a form of time travel. To carry it out the restorationist first has to understand the historic system he or she is trying to restore, and then must understand the various influences that have brought about change in order to reverse them. In some cases the lessons involved may be trivial or obvious. But they may also be subtle and may lead to a more complete comprehension of the system and its history; a convenient example is the rediscovery of the role of fire in the ecology of tall-grass prairies, which emerged from early attempts to restore these systems during the 1930s and 1940s.[10] In either case, the process involves the revisiting of history and the acknowledgment, at a practical level, of its implications for the present. Thus restoration is an exploration of change and its implications, and one of its lessons is the cost of change, as well as the crucial distinction between change that is reversible and change that is not.

Restoration explores history, but it also explores the slower rhythms of prehistory and cultural evolution. The restorationist not only attempts to reverse history but also to a certain extent recapitulates the major phases of cultural evolution, from hunting and gathering, to gardening and farming, to science. All the varieties of human experience of nature are repeated. The restorationist approaches a species of plant, for example, first as a gatherer, with an economic motive and a sense of appreciation for the plenitude of nature, not as "other" exactly, but as "given"; then as a gardener or nurturer of nature, who repays what he or she takes in kind as well as in gratitude; and finally as a scientist, who observes and manipulates nature in order to satisfy curiosity, and gives back to the world the gift of its greater self-awareness.

In this way the restorationist may travel back 10,000 years in a single afternoon. Of course to benefit fully we would like to know more about the classic relationships with nature that the restorationist revisits. For this we will be depending on anthropologists for precise descriptions of the subjective experience of nature characteristic of other cultures, but from what I have been able to discover so far, anthropology has little to say on this point. Anthropologists seem to have concentrated almost exclusively on the objective aspects of the nature-culture relationship—on calories and foraging patterns and the like, which are useful but only part of the information that we need. Perhaps the task of restoration will challenge anthropology as it has challenged ecology, and maybe restoration itself will allow us to test directly ideas about the subjective experience of nature by reducing these ideas to practice.

5. *Change and adaptation.* The reason underlying all so-called environmental problems, and in fact the general human sense of alienation from nature, is simply the speed of cultural change. Culture, like nature, evolves; but while the rest of nature evolves slowly, stuck, as it were, in the old, slow lane of chemical-based Darwinian evolution, cultural change has shifted into a computerized mode faster by many orders of magnitude than most of the ecological or evolutionary changes we see—or can barely see—going on in the world around us. Thus culture is always diverging from nature, and at increasingly higher speed as the rate of cultural change accelerates.

This causes considerable despair within the old environmental paradigm, with its defensive posture toward the conservation of natural areas. The problem, however, is not that change within the human community is necessarily inimical to the classic landscape, but that environmentalism, in its necessary defense of nature, has stressed protection from human influence and has by and large failed to come to grips with the problem of human interaction with nature. As a result its whole agenda, based on the idea of minimizing impact, becomes less and less tenable as human influence on wild nature becomes more pervasive and exotic. Because the fundamental problem is not influence, which is inevitable, but a failure to acknowledge this influence and a tendency to wish it away, the solution is not more protection and the erecting of higher and higher fences in a fruitless attempt to isolate nature from culture, but a program that frankly recognizes human influence on the natural landscape and then sets out to compensate for it.

Restoration does precisely this. The salient point is that, whereas environmentalism has tended toward a kind of idealism in its conception of nature, restoration is relentlessly pragmatic. It asks not how nature may be kept pure and uncontaminated but rather just how it is actually being affected by human activities, and how this influence can be reversed. What is involved is a continual dialogue rather than a program, paralleling in our dealings with the biotic community the dialogue that sustains a democratic society and makes it adaptable to change. The restoration-based paradigm reenters nature from the vantage point of any kind of culture and works out a new relationship in practical and psychological terms as change continues and as a culture diverges further and further from its native landscape.

6. *Celebration*. Environmentalism is a complex movement, embodying a wide variety of attitudes and ideas, but I think it is fair to say that the environmentalism of the past generation has generally not been optimistic about the prospects for a positive relationship with wild nature. This follows from the assumption that humans stand somehow outside nature, and that nature is therefore irreversibly compromised by the influence of culture.

There are valid reasons for this attitude. Culture is encroaching on nature nearly everywhere, and threatens both the biotic richness and the normal functioning of the biosphere. Yet it seems obvious that as the fate and well-being of the biosphere depend ultimately on us and our relationship with it, we must find out not only how to have a healthy ecological relationship with the world but also how to articulate and celebrate that relationship in a personally and socially effective manner.

Restorationists have discovered in recent years that the act of restoration can achieve and celebrate this relationship. An excellent example is the burning of the prairies in many areas of the Upper Midwest each year, usually in the spring. These prairies are in many ways the birthplace of the idea of ecological restoration. The dependence of these systems on fire was an early discovery of restorationists and one of their first fundamental contributions to the science of ecology. The burns are really the quintessential or emblematic act of prairie restoration. They have even become a rite of spring, eagerly anticipated by the growing number of "prairie people" involved in restoration efforts in the Midwest, and are often surrounded by a festive, joyful, atmosphere. Reflecting on this development, several years ago, Fred Turner put forward what I believe is a good explanation for it. It is not, he pointed out, just that burns are often spectacular, exciting events, tinged even with an element of danger, or that the fire is a powerful tool that can change the landscape drastically; the need of the prairie for fire dramatizes *its* dependence on *us*, and so liberates us from our position as naturalists or observers of the community into a role of real citizenship.[11]

The burning of the prairies is more than a process or a technology, it is an expressive act—and what it expresses is our membership in the land community. The implication is that we have a role here: we *belong* in this community, and so perhaps we belong on this planet after all. This, quite simply, is good news that makes people happy.

The implications obviously go far beyond the conservation of the prairies, offering an escape from the excessive and unrelieved negativism that is a kind of occupational hazard of environmentalists. This new perspective is revealed in the response of Steve Packard, a restorationist with The Nature Conservancy in Illinois, to *The End of Nature*, by Bill McKibben. This recently published book is a classic variation on the theme of human alienation from nature and the hopelessness of our present situation. Very briefly, his theme is that nature is everything in the world except people and their works, and that because all nature—including the atmosphere, McKibben's special concern—has been touched and contaminated by human beings, nature has actually come to an end. In reading this gloomy and destructive book, I hope that the elaboration of the logical consequences of its initial premises will at least serve the purpose of emphasizing how desperate and paralyzing these hypotheses are. In one particularly ugly passage, McKibben predicts, and in a sense even offers a justification for, a growing despair over the future of nature: "The end of nature," he writes, "probably also makes us reluctant to attach ourselves to its remnants for the same reason that we don't usually choose new friends from among the terminally ill."[12] Packard, reflecting on his experience with the prairies and oak openings of the Chicago area—terminally ill ecosystems if there ever were any—replies simply, "Our experience is the opposite. Unprecedented numbers of people are becoming passionately involved with the environment. It's an honor to be among the first to have a nurturing relationship with wild nature."[13]

Packard knows what he is talking about. He has direct experience as a pioneer restorationist and as the leader of a growing army of restorationists—now numbering over 4,000—who are reversing more than a century of deterioration in the Forest Preserve system of Chicago, rescuing the preserves from preservation, as it were, and bringing them back to nature. The work of Packard and others like him now points toward what I believe will prove to be most important about ecological restoration: its value not just as a process or a technology or a strategy for conserving bits and pieces of the natural landscape, but its significance as a performing art and as the basis for a new ritual tradition for mediating the relationship between nature and culture.[14]

This brings us to a crucial point in the development of the restoration-based environmental paradigm—the role of performance or ritual in mediating the relationship between nature and culture. This aspect of restoration has remained invisible to environmentalism for at least two reasons. First, in its preoccupation with nature as object, environmentalism has been concerned exclusively with the products of restoration (the restored communities themselves and their quality) and has paid little if any attention to the process of restoration and its implications for people, both those carrying out the restoration and those looking on—the audience, as it were. Environmentalism has then missed restoration's value as a way of reentering nature. The second reason is environmentalism's blindness to the performative or expressive aspect of restoration—to what might be called its ritual value—and to the crucial role of ritual in mediating relationships.[15]

Of course this is not peculiar to environmentalism but rather is characteristic of our entire society, with its reduced sense of the efficacy of ritual. Perhaps this is another "root" of the so-called environmental crisis that has developed in the West since the sixteenth and seventeenth centuries. Surely the emergence of science, and later the technologies based on it, played a role by increasing the distance between nature and culture. Even as that gap widened, however, the Reformation mounted an explicit attack on symbolism and ritual, and largely did away with the ritual traditions that human beings had always depended on in their contact with nature. The result, it may be, was a worldview within which real union with nature is impossible.

This blindness to the performative experience and its implications may be understandable in historic terms, but it is a deficiency in environmental thought and could prove to be fatal. Certainly it is at the root of much of the debilitating pessimism that environmentalism generates, because perhaps, from its own, puritan point of view, environmentalism has been right. A fully satisfactory relationship with nature actually may be impossible, and accessible only through recourse to another dimension, that of performance, ritual, and make-believe.

The underlying mistake here may be the perception that indigenous cultures are "natural" people who live more or less unself-consciously in harmony with nature. (Perhaps this is why we have traditionally put them, along with evidence of their often-impressive technologies and other cultural achievements, in our natural-history museums.)This seems to be a fairly widespread notion within environmentalism, where it serves as a kind of ideal and as the foundation for much thinking about the proper relationship between humans and the rest of nature. This view is by no means universally accepted by anthropologists, however. Indeed, it is my impression that most anthropologists see in all cultures evidence of a tension between nature and culture, which is then mediated or dealt with in various ways that to a considerable extent define the culture and lend it its distinctive characteristics. On this ground, then, I put forward the following premises as the basis for a new paradigm for the relationship between nature and culture:

1. Though ourselves the products of nature, and in this sense natural, we do differ in certain fundamental ways from the rest of nature, notably with respect to our level of self-awareness. Thus we may be citizens of the world, but we are not "plain citizens," and any attempt to overlook this is simply wallpapering over a major feature in the structure of the world, and is bound to have unfortunate consequences. People have never regarded themselves as "plain citizens" of the world; instead they have always—at least since the development of language—distinguished between nature and culture and have felt a measure of tension between themselves and the rest of nature. Moreover, though it may vary in intensity, this tension is irreducible. It cannot be avoided simply by living in a simpler or more primitive way, "closer to nature." It is part and parcel of being human; it comes with our genes.

2. Although this tension cannot be resolved in purely literal terms, it can at least be dealt with in a psychologically effective way through performance and ritual. This, then, is one of the functions of ritual, and humans have used ritual techniques from time immemorial to mediate their relationship with nature.

3. The process of ecological restoration provides an ideal basis for the development of a modern system of rituals for negotiating our relationship with the rest of nature.

This, then, is the outline of a new environmental paradigm based on a sense of the crucial role of ritual in any satisfactory relationship between ourselves and the rest of nature, and on the observation that the act of restoration provides an excellent foundation for the development of a new ritual tradition. I should stress that what I have in mind is not simply the addition of performative techniques such as music, poetry, and so on, *to* the process of restoration, but a conception of restoration itself as both an effective process and an expressive act. The idea is not merely to *decorate* restoration, but to develop it to enhance its expressive power.

This conception is at the heart of Earthkeeping, a new program being developed by the University of Wisconsin–Madison Arboretum and the Society for Ecological Restoration to provide opportunities for people to participate in restoration efforts at selected sites as a way of learning about a healthy relationship with nature. In my view this is a step toward the emergence of restoration as a major cultural event, comparable with other social rituals such as elections, sporting events, festivals, and holidays—and toward Aldo Leopold's "civilized society" living not in harmony, but at least in an ongoing dialogue with the natural landscape.

CONTEXT, DEFINITIONS, AND CLARIFICATIONS

The ideas presented in this essay are the result of some fifteen years of reflections and discussions concerning the development of a collection of restored ecological communities at the University of Wisconsin–Madison Arboretum. This project was undertaken in 1934, under the leadership of a handful of ecologists and conservationists that included Aldo Leopold, Ted Sperry, John Curtis, and Henry Greene. It was a pioneering project from the first, and today the resulting collection of restored prairies and forests is considered the oldest and most extensive such collection in the world. It is widely regarded as a model for the idea of ecological restoration in its strictest sense, and has served as an inspiration—and in some cases even as a source of seed—for numerous projects at other locations.

Today the Arboretum has begun to serve as a symbol of ecological restoration, and in my view it will one day rank with places such as Walden Pond or Yosemite National Park as a landmark in the development of the modern environmental sensibility. For purposes of this essay, what is important about the Arboretum is its demonstration of the methods and objectives of ecological restoration. A good example is the John T. Curtis Prairie, a restored tall-grass prairie that was the first major restoration project at the Arboretum and which now occupies 64 acres in the center of the 1,280-acre teaching and research facility.

Survey records from the 1830s indicate that this site was covered by a mixture of tall-grass prairie at the time of European settlement, but when the Arboretum was dedicated in 1934 the original vegetation had been nearly eliminated and the site had been under cultivation for about three-quarters of a century. Restoration at that time meant attempting to return the historic grassland vegetation to a site occupied mainly by exotic grasses and weeds. Work began on a large scale in 1936 under the supervision of Dr. Theodore Sperry and has continued at varying levels of intensity ever since. Curtis Prairie and the slightly younger and smaller Henry Greene Prairie about one-half mile away are currently regarded as the oldest restored prairies, and quite possibly are the oldest restored ecosystems anywhere. They have been the subject of many scientific studies over the years, and some parts of these prairies, especially of Greene Prairie, are considered among the highest-quality replicas of a natural prairie ever achieved.

The effort that led to the restoration of these prairies, though novel in certain respects, was not altogether unprecedented. It drew in part from related activities in areas as diverse as forestry, landscape design, and wildlife management. What set it apart from earlier efforts, however, was the commitment not just to manage the land, or even to rehabilitate it in a general sense, but to re-create, deliberately, a faithful replica of a historic ecosystem.

This activity, with this explicit purpose, is what is meant here by the term "ecological restoration." It is important to keep in mind that this can be carried on at several levels. The work at the Arboretum was relatively dramatic because it involved an attempt to replace virtually an entire ecosystem wholesale on a site from which it had been almost completely removed. This makes for a clear illustration of the principle behind ecological restoration, which is simply the active attempt to compensate for human influence on an ecological system in order to return the system to its historic condition. This continual effort to sustain the system against the pressure of our own influence makes restoration such a powerful tool for exploring and defining our relationship with the system, and for achieving what might be called an ecological definition of who we are — that is, a definition written in terms of our impact on other species and ecosystems.

The degrees of influence involved in this process may vary enormously, from very great, as with the University of Wisconsin–Madison Arboretum, to subtle. When the influence is subtle many prefer the term "management" to "restoration." But there is no clear distinction between restoration and management in these senses: they are simply parts of a continuum. In my view it is important to reject the false distinction between them and to refer to activities across the entire continuum as restoration because this explicitly acknowledges the role of the human in the process and opens it to the subjective benefits explored in this essay. To do otherwise is to avoid the responsibility of biotic citizenship or perhaps to reserve this responsibility and the satisfactions and benefits associated with it to a professional elite, an approach to conservation that I believe will inevitably fail in a democratic society.

NOTES

1. Interestingly, Leopold used this phrase in a speech at the dedication of the University of Wisconsin-Madison Arboretum in which he outlined the then-novel plan for a large-scale ecological restoration project on the property. Two versions of this speech survive in written form. The longer, which contains this phrase, was printed in a booklet commemorating the fiftieth anniversary of the Arboretum in 1984. The other, shorter version first appeared in *Parks and Recreation* magazine and is included in a recently published anthology of Leopold's writings; see S. L. Flader, and J. B. Callicott, eds., *The River of the Mother of God and Other Essays by Aldo Leopold* (Madison: University of Wisconsin Press, 1991), 209-11. The versions are quite different, and the latter refers to "harmonious relationship" instead of "mutually beneficial relationship."

2. The phrase is from the Wilderness Act of 1964, one of the early achievements of the modern environmental movement, and arguably one of its most characteristic, at least as far as the natural landscape is concerned.

3. Henry D. Thoreau, *The Journal of Henry D. Thoreau*, 2 vols., Bradford Torrey and Francis H. Allen, eds. (New York: Dover, 1962), 1:53 (entry for June 16, 1840).

4. See S. Packard, "Just a Few Oddball Species: Restoration and the Rediscovery of the Tallgrass Savanna," *Restoration & Management Notes* 6, no. 1 (1988): 13-22. For another example from the West Coast, see Rich Reiner and Tom Griggs, "Nature Conservancy Undertakes Riparian Restoration Projects in California," *Restoration & Management Notes* 7, no. 1 (1989): 3-8. These early projects have demonstrated the value of restoration by volunteers to the natural-area conservation efforts of an organization such as The Nature Conservancy. Though novel when first undertaken within the past half-dozen years, these projects are now regarded as models, and such work is expected to play a central role in a new plan for the Conservancy that some insiders have called "revolutionary."

5. Loren Eiseley, *The Invisible Pyramid* (New York: Charles Scribner's Sons, 1970). The references are to passages from chapter 7, "The Last Magician."

6. The phrase is from the opening paragraph of the chapter on "Solitude" in *Walden*.

7. William R. Jordan III, Michael E. Gilpin, and John D. Aber, eds., *Restoration Ecology: A Synthetic Approach to Ecological Research* (Cambridge: Cambridge University Press, 1987), especially the introductory chapter.

8. Mircea Eliade, *The Myth of the Eternal Return* (Princeton: Princeton University Press, 1954).

9. The use of this term is borrowed from Paul Shepard, *The Tender Carnivore and the Sacred Game* (New York: Charles Scribner's Sons, 1973).

10. See J. T. Curtis and M. L. Partch, "Effect of Fire on the Competition between Blue Grass and Certain Prairie Plants," *American Midland Naturalist* 39, no. 2 (1948): 437-43.

11. Frederick Turner, "A Field Guide to the Synthetic Landscape: Toward a New Environmental Ethic," *Harper's Magazine* 276 (April 1988): 49-55.

12. Bill McKibben, *The End of Nature* (New York: Random House, 1989), 211.

13. Steve Packard, "No End to Nature," *Restoration & Management Notes* 8, no. 2 (1990): 72.

14. W. R. Jordan III, "A New Paradigm," *Restoration & Management Notes* 9, no. 2 (1991): 64-65. See also my editorials in other issues of *Restoration & Management Notes*, including 5, no. 1; 7, no. 1; 7, no. 2; and 10, no. 2.

15. An interesting example occurs in the report on the management of the national parks prepared by a commission headed by Starker Leopold in 1963. While prescribing what is essentially a program of ongoing restoration for the parks, even describing them in theatrical terms as vignettes of the presettlement landscape maintained to create an illusion of original wilderness, the report insists that the work of restoration itself be kept out of sight—backstage, as it were. This is a classic expression of the conception, characteristic of modern environmentalism and distinguishing it from conservation movements earlier in this century, of nature as a collection of objects in the landscape—in fact, literally an "environment." Though the Leopold report sees restoration as a performance in a sense, its interest at least so far as the public is concerned is exclusively in the product of restoration: the "finished" ecosystem as an object in the landscape (that is, more an art like painting or sculpture, with their emphasis on the creation of concrete artifacts, than like the performing arts, with their ecology-like emphasis on process and relationship). This, however, deprives the public of the experience of restoration—either as audience or as participant, and excludes the people from the very process that defines our relationship with nature. The result is an illusion of nature as pristine and apart. Our relationship with it then becomes the responsibility of a corps of experts working behind the scenes. The elitism implicit in this formulation, though obviously unintended, would in my view prove fatal to conservation in a democratic society.

The Invented Landscape

Frederick Turner

Environmentalism Classified

Four main currents of thought can be detected within the ecology movement at the present time: the conservationist, the preservationist, the restorationist, and what might be called the inventionist. The first two are very familiar, the third less so, and the fourth is so unrecognized that it does not have a fixed name — perhaps this book will help to give it one.

Conservation sees nature as a vast resource, physical and spiritual, that must be wisely husbanded so that it may continue to yield a rich harvest for human beings. Preservation sees nature as of intrinsic value, the greater for being untouched by humankind, and seeks to keep it inviolate and unpolluted. Restoration, which I believe is presently the most intellectually exciting organized field of ecological philosophy, seeks to reconstruct classical ecosystems, and is based on an assumption that to do so is not only possible, given nature's own easygoing and flexible standards, but also an important part of the human role within nature.

The restorationist philosophy is not necessarily at odds with the goals of conservation or preservation. But its purpose is different from conservation inasmuch as, for the restorationist, nature is of intrinsic worth as well as being valuable for human uses. Though they agree on nature's intrinsic value, the restorationist's definition of nature is different from that of the preservationist: restoration holds that a restored landscape is no less natural — and may even be more natural in some senses — than an "untouched" one. All landscapes, as restorationists know by experience, are always already touched—if not by earlier human beings, then by the world-transforming activities of the species that constitute them, or by other natural interventions such as volcanic eruptions and meteor strikes. Nature is the process of everything interfering with—touching—everything else. Human beings, another natural species, are part of this process of irreversible change, but it is quite possible that termites have transformed the planet at least as much as have humans (they have had more time to do so, however). Because the evolutionary mechanism of life is reproduction with differences in each generation (in other words, a system of imperfect copying), there is nothing unnatural about a copy of prairie, if accurate enough. William Jordan's essay in this volume is, I believe, the best and most authoritative statement of the restorationist philosophy to date.

This essay has a different task: to sketch the outlines of a nascent ecological philosophy, one that for lack of a better name we have dubbed "inventionist." This rather ugly word has a hidden ambiguity or paradox, which, I hope, may be the sign of an inner beauty. The Latin verb *invenire* means "to come upon"; thus the surface meaning of the word invention, which brashly announces its radical originality and heroic innovation, is counteracted and modulated by the etymological implication that in some sense the new thing that is invented is really an old thing that was there all the time, waiting for us to "come upon" it.

Inventionist ecology, to give a crude definition, maintains that it is both possible and desirable not only to conserve natural resources, preserve natural ecosystems, and restore natural landscapes, but also, when the occasion warrants and the knowledge is sufficient, to *create* new ecosystems, new landscapes, perhaps even new species. Inventionist ecology does not contradict the goals of conservation, preservation, and restoration: far from it. Conservation is simply rational common sense and the solid foundation for any human activity, though insufficient by itself either practically or philosophically. Preservation is necessary at least to maintain the library of genetic richness from which our work can grow; only when the librarians seek to stop the public from touching the books must we gently dissuade them. Restoration is totally in harmony with invention, and indeed invention cannot happen without the tools and understanding provided by restoration. Moreover, it could well be argued that restoration is a special form of invention, a form whose aesthetic, like that of Aristotelian drama, is imitative or mimetic.

Potentially at least, human civilization can be the restorer, propagator, and even creator of natural diversity, as well as its protector and preserver. This will come about as biological science and technology are perfected, and as the traditional ecological wisdoms embodied in ancient gardening and husbandry are rediscovered through the practice of ecological restoration, or "earthkeeping," as it is known. Thus we will go from being a net destroyer of biological information and a (larger) net creator of cultural information to being a net creator of both. The dream of inventionist ecology is that eventually we will propagate life into presently dead regions of the universe, and even assist in the development of entirely new species.

This formulation dictates the main questions of this essay. How can the creation of new landscapes and species be ethically justified? What occasions might warrant it? Are human beings capable of such an invention? How much knowledge is sufficient? What guides exist? Has it ever been done before?

In order to address these issues, first the rudiments of a comprehensive environmental philosophy must be developed. But environmentalism, dealing with the whole cosmos and the human place in it, is more than a philosophy; given the definition of the divine as ultimate concern, it is also a theology. Thus this essay will question the contemporary environmental theology, which is hostile to invention in such matters, and replace it with one that permits and even mandates inventionist ecology.

Tenets of Traditional Environmentalism

If asked to state the goal of the environmental movement, a contemporary participant would probably say something like, "to promote a sustainable relationship between human beings and Nature." How could one object to such a formulation? Yet hidden in it is a set of assumptions that may paradoxically lie at the root of our present environmental crisis.

There is a close resemblance between this stated goal and a much older idea from the rationalistic theology of the modern West: that the goal of the moral life is to promote a sustainable relationship between human beings and God. Just replace God with Nature, and much of the environmental movement looks very like a church, though without the disadvantage of being separated by law from the state.

A brief digression into theology may help to make this point clear. The God of Christian theism is eternal, transcendent, perfect, and unchanging. Obedience to Him is true happiness. We human beings are inherently fallen and wicked in part, and thus a caste of priests is necessary, themselves under the control of an ecclesiastical organization, to discipline the population in the correct beliefs and religious practices.

The Christian theism of the West was the heir of an older religious worldview, that of ancient Christianity, which was much closer to the naturalistic, pagan, and polytheistic religions of the ancient Mediterranean. Ancient Christianity was less abstract, less transcendent, more ritualized, and more performative than the modern theism that succeeded it. In that more ancient religion God, the angels, and the saints (and before that, the gods) were more present within the world, less perfect, less absolute in knowledge and power, more in process, more a story than a theology. The divine was *in* the world, not outside of it, and the divine made itself known in the forces of nature. Christianity has periodically regretted its departure from that ancient religious milieu, which was much more appropriate to the life and thoughts of Jesus Christ than the very philosophical transcendent theism of modern Christianity.

But the environmentalist ethic has in effect replaced God with Nature; and the God it replaced was the God of modern theism, the abstract, unchanging, emotionless, moralistic authority who dwells outside the universe and who, knowing all in advance, can never be surprised, or grow, or have a story. Very often the environmentalist's idea of nature retains these characteristics of the transcendent God. As the phrase goes, "It's not nice to fool Mother Nature." James Lovelock's Gaia hypothesis, which argues plausibly that in some sense the planet Earth is a sort of superorganism, perhaps like a giant polyp or colonial animal or coral reef, maintaining its own atmosphere, climate, and chemical environment, has supplied its more religious followers with a personal name for the new deity: Gaia. But the fact that the name and sex of the deity have changed does not mean that the new environmentalism is any less a religion.

I would like to make clear before I go any further that I believe that religion is a unique means of apprehending profound truths, values, and beauties in our experience that cannot be captured in any other way. I also am deeply sympathetic to many of the goals of contemporary environmentalism, and work for them myself. However, I believe that some fundamental errors in the new naturalistic religion can be corrected if we recognize the implications of a theology that places the divine within, rather than outside, the natural universe.

Many contemporary environmentalists would probably accept without much question certain assumptions that are essentially religious, though not yet recognized as such. These concepts already function in public policy, demonstrating possible exceptions to the separation of church and state, and would include the following—what we might call the unspoken principles of the ecological religion.

- The basic feature of Nature is homeostasis; there is a natural balance that is restored when it is disturbed, and a natural harmony. Nature in this view has an ideal state, which is perfect and should not be tampered with. (This is an unpacking of the central term "sustainable.")
- Happiness is doing the will of Nature. Human beings are evil and distorted creatures, filled with greed and the desire to dominate, an unnatural presence in the universe. If they are converted, however, then they will find true happiness in going back to nature, in humble service to the environment. (This feature of the environmental movement is much like the doctrine of original sin and salvation.)
- Happiness for human beings is fundamentally stasis, an unchanging and secure state in which the future is more or less predictable. Change, especially swift change, is evil.
- Humans are different and separate from, and subordinate to, a transcendent Nature. (Here the environmentalist religion ignores the central scientific principle of evolution, which treats humans as a part of nature, though contemporary environmentalism is quite happy to use scientific research if it seems to prove the point of human destructiveness.)
- Human beings are no better and no more important than any other species. (The Christian doctrine of the equality of souls before God has been translated into a doctrine of the equality of species before Nature.) In a sense, this principle contradicts the previous one, in which human beings are uniquely wicked among species, and separate from Nature. The contradiction can be partly resolved by saying that in a state of nature, human beings are just another animal species; where we went wrong—where we "fell"— was in thinking ourselves better than the others. But the problem remains: why shouldn't it be natural for us to think ourselves better—and why shouldn't it be true?!
- An (unelected) community of environmentally conscious, morally refined, sober, devout, humble, and self-denying ecological Brahmins should interpret to the masses the will of Nature and

direct them accordingly, chastising the merchant/industrial caste, humbling the warrior caste, and disciplining the farmer caste.

Critique of the Traditional Faith

This creed stands or falls on whether nature really is as its believers maintain—eternal, unchanging, and so on—and whether human beings are indeed as the theory says they are. As far as I can make out, there are four main problems with what we might call the "ecotheist" creed.

The first is that nature is not and has not ever been static and unchanging. It is easy to demonstrate that nature is a process of irreversible changes at every level of the microcosm and the macrocosm, both in the world of living organisms and in the world in general.

Let us start with the world of life. Since life appeared on Earth about four billion years ago, the number of species has pretty steadily increased, with occasional episodes of mass extinctions. One of those mass extinctions was at the Cretaceous/Tertiary boundary, when it is thought a giant meteorite hit the Earth. This cataclysm, swifter and more devastating than any human effect on the environment, actually resulted in the rise of many more species, such as the mammals, and in the long run brought about an enrichment of the ecosystem. It is not just the number of species that has increased over the lifetime of this planet, however; so too has the complexity, hierarchical organization, and neural development of the most highly organized species. Though many very primitive and simple species still exist, the range of development among the species is now much greater than in the past. Thus the ecotheist principle that nature is homeostatic is false for living creatures.

It is false also for the inanimate parts of nature. The universe at large has experienced a steady and irreversible increase in several crucial measures, including its size (it has been expanding since the Big Bang) and its age. The thermodynamic entropy of the universe has increased; that is, energy has distributed itself into smaller and smaller packets, with less and less difference between them. The complexity, sensitivity, and degree of freedom of even inanimate objects has increased: the definition of time required to describe a photon (which preexisted all the more complex forms of matter) is much simpler than that required to describe an atom, a molecule, a crystal, a bacterium, an animal, or a human being, each of which inhabits a more complex temporal environment than its predecessors. The amount of feedback or self-reference in the universe has increased irreversibly. The amount of information in the universe has increased. The universe has cooled down irreversibly and will continue to do so. Even the *rate* of all these changes has been changing irreversibly, and for some of them also unpredictably.

Parts of nature do not, indeed, change very much, but these are the least alive parts—neutrinos, for instance. Other parts of nature are governed by negative feedback loops that are cyclical and homeostatic, and if permitted to do so restore their balance when they are disturbed, like a thermostat regulating the temperature of a house. But still other parts are involved in positive feedback processes that are irreversible, catastrophic to their predecessors, and often wildly original and creative. If I put a live microphone in front of its speaker, any small sound can be magnified by positive feedback into an uncontrolled shriek. Nature does not avoid such situations: they can be found in the sudden crystallization of a freeze, in the wild multiplication of bacteria in a dead animal, in the recolonization of volcanic wastelands, and in the famous "butterfly effect," whereby the turbulence caused by a butterfly's wingbeat can, in the right humid circumstances, be amplified into a whirlwind, a tropical storm, and a hurricane.

Life is full of such processes; collectively they are called evolution. The air of the Earth was the result of catastrophic changes brought about by the evolution of photosynthetic organisms, which gradually "poisoned" the early terrestrial atmosphere with oxygen so that the ancient prokaryotic inhabitants of the planet could no longer survive and had to give up their places to new, more adaptable organisms that were our ancestors. The natural disasters that have punctuated the Earth's history, whether the result of gigantic volcanoes or meteor strikes, may well have speeded up the process of evolution. It is entirely possible that the increase in carbon dioxide in our atmosphere will cause an increase in plant metabolism and world precipitation, and lead to an age of unparalleled natural fertility and species proliferation rather than an age of ecological catastrophe. Only the faulty assumption that any change is unnatural makes us conclude that the greenhouse effect will be bad for the planet.

Thus the idea of "sustainability" and general homeostasis is a profoundly unnatural goal. The universe does not, except in certain temporary periods and places, sustain or maintain: it changes, improves, complexifies, sometimes destroys. Sexual reproduction, to take a good example, consists of a sophisticated and powerful mechanism to ensure that the genetic inheritance of a species changes irreversibly: it is a system to subvert and disrupt sustainability and maintainability. Organisms that clone themselves, and play it safe, are opting for sustainability; but it is the more advanced, sexually reproducing organisms, which allow their genetic codes to be reshuffled every generation, that have driven evolution. We human beings may still want the security of sustainability, but we should not invoke the authority of Nature to bring it about.

The philosophical error of assuming that nature is essentially unchanging has led to actual damage to ecosystems managed by well-meaning believers in natural homeostasis. The ecological scientist Daniel Botkin cites many examples in his book *Discordant Harmonies*, including the reduction of huge fertile nature preserves in Africa to semideserts by exploding elephant populations, managed by preservationist true believers who could not face the necessity to cull the elephant population. Another example is the way in which various different but equally erroneous notions of natural balance led to the recent destruction of large areas of Yellowstone National Park. Wiser managers of prairie preserves in the Midwest, like those William Jordan describes, have learned that a good prairie needs to be burned from time to time; fire, the destructive element, must be wielded by human beings if we would keep nature in the same state. A similar philosophical puzzle confronts the managers of Niagara Falls, who may need to turn the falls off for a time, while the natural undermining of the rocky scarp over which it flows is repaired with concrete!

Another problem with the idea of sustainability derives from the axiom that humans are happiest when in harmony with an unchanging natural world. But it is surely naive to assume that human happiness can consist in stasis and obedience. Elementary psychology tells us that we are sensation-seeking animals, and that our brains work upon a principle of habituation and fatigue; if we encounter the same stimulus for a period of time (say, a ticking clock) we very quickly discount it and cease to notice it. It is our nature—and the nature of many higher animals—to seek out new stimuli. Stasis is thus sensory deprivation, which is the most subtle and severe form of torture we know. Human happiness cannot consist in lack of change. It used to be thought that so-called primitive tribal societies lived in an unchanging and harmonious state of peace, without history; but a wiser anthropology has shown us that this view is in error. Such societies do change spontaneously every generation, in their rituals, beliefs, fashions, even in their kinship system. My father, the anthropologist Victor Turner, could not at first understand why the Ndembu people of Central Africa would laugh when he asked about their clan system. This kinship structure had been recorded by an earlier anthropologist, Charlie White, forty years earlier, as being the key to Ndembu society. Finally my father discovered that the Ndembu had changed to a different kinship system twenty years before his study, and the clan was as obsolete as a Victorian shovel hat. Why had they changed? Boredom.

A more sophisticated version of the environmentalist position has recently abandoned the usual praise of natural homeostasis and asserted that only in the wild, the wilderness areas, can Nature find its true freedom to evolve and develop, in its naturally irreversible and unpredictable way. This argument reverses the usual complaint against human culture, that it changes things too fast, and declares that human culture, by taming and domesticating Nature, robs it of its creative powers of metamorphosis. This is a serious and interesting position, and clearly holds some truth that should be incorporated into the wiser environmentalism that is the goal of this essay. Its imagination is limited, however. For is not the disturbing, horrifying, unpredictable, dangerous, and protean character of human culture and technology, like Max in Maurice Sendak's fable, the wildest thing of all, the true wilderness that lies beyond the edge of the tamer, more serene, and self-maintaining fields of the terrestrial ecology? Is not human culture, as compared to the rest of nature, like a sexually reproducing species surrounded by an ecology of clones?

This reflection leads us to the second major objection to the ecotheist cliché that is serving us so usefully as a straw man: the distinction it draws between the human and the natural is patently false.

We are descended in a direct evolutionary line from natural animal species, and are ourselves a natural species. Our nature, certainly optimistic, transformative, activist, and bent on propagating itself, is not unlike that of other species, only more so. We are what nature has always been trying to be, so to speak. Nor can it be objected that the *speed* at which we transform ourselves and the world around us is unnatural. Higher animals evolve faster than do more primitive organisms, as these in turn do so faster than nonliving systems. If we take flexibility, complexity, hierarchical organization, and self-referentiality as the measure, we may define nature as acceleration. For us to slow down would, if we take nature *to date* as the model of what is natural, be unnatural.

Likewise, the greater self-awareness of human beings compared to other species must be seen in a context in which the other higher animals are more self-aware than lower animals and plants, life is more self-aware than matter (in that it contains a record and blueprint of its own structure in the sequence of its DNA), and matter is more "self-aware" than energy, as matter is energy bound to itself rather than flying off radially from itself at the speed of light. Thus humans also appear to be the closest approximation nature has found to its own direction and tendency. If human beings are what nature comes up with given the freest play of development and the richest and stablest environment of complexity and available energy, then we might rightly assume that if we want to know what nature is *really* like, we should look at ourselves.

Not that I am necessarily advocating a continuous indiscriminate acceleration of our activities; but any moratorium we call cannot honestly be claimed to be in the name of Nature. Nor am I denying the human fact of evil actions, both against other human beings and against the rest of nature. In the old religion that serves as the tacit model for some aspects of current environmentalism, that evil was explained by the Fall. But in that religion there was also a wisdom that called the Fall a *felix culpa*, a happy fault. The good of knowledge, freedom, and possibility of divine redemption came with the darker lapsarian consequences of death, guilt, and the propensity to sin. My claim is that nature itself, like ourselves, is fallen, is falling, and has always been falling, outward into the future from the initial explosion of the Big Bang; onward into more and more conscious, beautiful, tragic, complex, and conflicted forms of existence, away from the divine simplicities and stupor of the primal energy field.

Thus for good and ill we are in solidarity with the rest of nature; and though there may be a vaguely good moral intention to the injunction that humankind should live in harmony with nature, the idea is essentially incoherent. Perhaps one might take it as a slightly stretched metaphor, as a doctor might advise living in harmony with one's kidney or liver, or that our brain should live in harmony with our body.

The third objection that might be raised against the "ecotheist" position is that the notion of universal species equality is, on analysis, inconsistent. It should not really be necessary to argue whether a human being or an AIDS virus is more valuable, but we are forced to such measures by the assertions of some of the more extreme deep ecologists, who question our right to consider ourselves more important and valuable than other species, and thus to affect their destiny. This point looks reasonable if we think about whales, bears, and other animals with whom we can easily identify, but it leads to deep absurdities. Is a bear committing an ecological crime by scratching itself, and so killing several million of its own skin cells, which have just as much right to exist as the bear? Why indeed should such theorists of universal species equality draw the line at living organisms? Are they not guilty of vitalist chauvinism, in not giving equal rights to crystals, clumps of amorphous matter, atoms, photons? To be consistent with the doctrine of equality, we should not consider energy any less valuable than matter. Yet energy is often locked up in matter—should we not "liberate" the billions of photons tyrannically imprisoned in the matter of the Earth, by blowing it up with nuclear bombs? The equality argument is plainly ridiculous.

One of the fundamental principles of nature is hierarchy, as the food chain, the delegation of control down the nervous system, and the branched subordination of functions within a given living organism amply illustrate. Though interdependence is another basic principle, it does not imply equality. A fox will bite off its own leg to escape a trap, choosing against the living cells of its leg and for the rest of its body. The brain and the kidney may be interdependent, but any deep ecologist would rightly insist that his own kidney is subordinate to and less valuable than his brain, and would rightly demand of his surgeon that she sacrifice a kidney if necessary to save even a small piece of his brain.

Although the combination of ideas can cause cognitive dissonance in some minds, we humans are both part of nature, and superior to and more valuable than any other part; at the same time we are essentially dependent on the rest of nature, and the loss of any of its unique and beautiful forms is an absolute loss to us.

Our final objection is political. Any attempt by an ecological elite to impose Brahmanic control over the masses—over the merchant, warrior, and farmer castes, so to speak—is doomed to failure. In Eastern Europe the Communist party was just such an enlightened and refined elite, and indeed, as we are finding, it did help to keep down such atavistic tendencies as ethnic hatred. But all over the world those masses have sacrificed themselves, suffered, and died, to escape or overthrow this new form of priestly control. Moreover, the best-intentioned Brahmanic bureaucracy can be ecologically disastrous, as we have seen in the devastation left by iron-curtain industry in Poland and East Germany. Even the mild forms of state control over natural resources that we in the West have instituted can, especially when energized by ideology and the urge toward bureaucratic survival, paradoxically produce serious ecological damage, as in the case of the Forest Service and the public rangelands. There are more miles of Forest Service roads (which determine the service's budget) than of interstate highways.

To sum up, the formula "a sustainable relationship between human beings and Nature" is profoundly misleading. Nature does not sustain, but changes cumulatively, sometimes preserving earlier states while inventing new ones, and integrating old and new together in a more reflexive and self-observing way. There is no "between" the human and the natural, unless there can be a special relationship, not between one thing and another, but between the most characteristic and quintessential part of a whole, and the whole of which it is the privileged part—privileged because it is the most developed product of its own evolutionary process. Human beings are not equal to, but superior to, other species. The complete injunction to the sustainable relationship as formulated is politically impossible to enforce and often counterproductive when enforced.

A New Environmental Theology

Nevertheless, there is a residual wisdom in the call for the sustainable relationship, and it might be worth our while to try to reformulate and rescue this goal by providing a sounder philosophical (and theological) basis for it. Cutting our theology to fit our environmental ethics may seem odd, but, in a playful spirit, let us do so anyway. Perhaps afterward we can see whether the result makes sense philosophically and morally, and discover if it has any significant continuity with the best of our religious traditions. If contemporary environmentalism is a new religion based upon a faulty idea of nature, how would it differ if we corrected its errors? What would a religion faithful to natural science look like?

Let us begin by following the Gaia hypothesis in its theological implication that the divine is present within the world, not detached from it. This is not necessarily to adopt a pantheist position—that is, that the world *is* God. If, by analogy, we assert that the mind and soul are present in a brain and body rather than detachable from them, we are not committed to believing that the mind and soul are only the brain and body. Thus the first axiom of a natural theology would be: *The divine is in nature.*

If the divine is *in* nature, how might we discover the nature of the divine? Surely by examining and listening to nature itself, just as we find out about man's or woman's personality by examining what they physically do and listening to what they physically say. That is, we should pay attention to the process, the *story*, of nature, if we wish to know its divine soul. Nature, as we have already seen, includes us as its acme and quintessence, so we must look especially at ourselves, the most characteristic part of God's natural body. The way we find out, the process of knowing, the attempt to come to know the story of things, is called science. Our second axiom might be: *We know the divine by means of the scientific understanding of nature and ourselves.*

A story is an irreversible process of events that are unpredictable beforehand but apparently inevitable and obvious once they have happened. If you are reading a good novel, the pleasure is partly that the next twist in the story can't be predicted—which is why you want to see how it comes out, whether the butler did it, for instance. Of course afterward there must be a good explanation; we should be able to say, "Why didn't I see it? The answer was obvious." Obvious after, but not before.

The possibility of story implies that time is asymmetrical, that truth can be different prospectively from what it is retrospectively. There are fixed truths, such as the laws of gravitation and thermodynamics, or we would have no points of reference by which to know. The newly emergent truths, however, include most of what we consider valuable, good, and beautiful: all the exquisite forms of matter, life, and mind that have evolved over the history of the universe. If nature has no story, then we can conclude that the divine being is fixed and eternal, forever unsurprised and undisturbed. But nature manifestly and emphatically does have has a story (or many stories), so we must conclude that the divine being has one, too. The third axiom of our natural theology would be: *The divine changes; the divine has a story.*

If we examine nature and ourselves we discover underlying unities in the variety of things (the mathematical forms, the constants of physics) but also that nature is an evolutionary drama, a competitive/cooperative dialogue among its parts, species, levels, and principles. If nature is the body of the divine, we may infer a fourth axiom: *The divine is both one and many.* It is one in its most remote, abstract, timeless, impersonal, simpleminded, and passive aspects, and many in its most immediate, concrete, changing, personal, intelligent, and active aspects. In deference to our own monotheistic tradition I shall from here on refer to the divine as "God," but with the understanding that the polytheistic and pluralist description of the divine as "the gods" is also intellectually attractive.

The transformations of this natural god of change are not exclusively random, reversible, and meaningless; as we have seen, the evolution of the universe is progressive, irrevocable, and dramatically meaningful. There is a one-way process of increasing feedback, reflexivity, self-organization, and freedom as the world evolves. Elementary particles have polarity but no shape. Atoms, more complex and self-referential than particles, have simple geometrical shapes that are symmetrical in many dimensions. With molecules, which could not exist until the universe had cooled enough to permit them, we see the first asymmetrical shapes and the birth of individuality. Molecules have complex feedback systems, many degrees of freedom, and the capacity to organize in periodic structures such as crystals. Living organisms are even more asymmetrical, free, and capable of organization, and they contain a recording of their own structure in the DNA language. Mind continues this story into the most complex forms of consciousness, self-determination, and communication. Thus the fifth axiom: *The story of God is one of increasing individuality, meaning, and freedom.* Progress is not a human invention, but a divine one.

If the universe is God's body, then we—and by "we" I mean all the intelligent species in the universe—are the most sensitive, most aware, most self-organizing of its parts. Though we are not the whole, we are that which increasingly has some knowledge of and control over the whole. The most sensitive and aware and controlling parts of a living body are its nervous system. Thus the sixth axiom: *We are the nervous system of God.*

This nervous system is still very rudimentary, and has penetrated and innervated only a tiny portion of the universe to date. It is like the nervous system of an unborn child. We stand at the first trembling moment of the history of the universe, the flash of a dawn that is a mere twenty billion years old, the beginning of a day that lasts ten trillion years. The universe is still only in its gestation; it is not yet fully developed. We bear some of the responsibility to complete that development, to increase our awareness and control over the rest of the universe, to extend the nerves of science and art into the inanimate and insentient parts of the world. Thus a seventh axiom: *God is still only a fetus.* Nature has not died, as some recent commentators have complained. It is only now awakening, and we are its eyes, its ears, and its tongue. From this follows an eighth axiom: *We serve God by helping him or her toward greater self-awareness.*

As organisms evolve, they develop more and more complex chemical, electrical, and mechanical systems, known as bodies, in order to control and be controlled by their environment—to act and to sense. All bodies are prostheses, that is, the matter of which they are made is not at first part of the living organism itself, but pressed artificially into service by that organism. For instance, the carbon atoms that my body uses to construct its protein and enzyme factories are exactly the same as they were before I commandeered them by eating them in my asparagus. Likewise, the coat of tiny sticks and bits of gravel that a caddis-worm constructs for itself is part of its body, though not strictly alive. The body of a termite colony includes its nest, that marvelously air-conditioned residence containing nurseries, storehouses, factories, and farms. A beaver colony is a similar example. The nest of the male blue satin bowerbird is not even used as a nest at all, but as a piece of advertising, a communication device to persuade a female bowerbird to mate. Yet in a strong sense that nest is part of its body. Plants and animals use probes, crutches, shelters, tools, vehicles, weapons, and other prostheses that do not need to be directly connected to their flesh or nerves, but which are essential parts of their bodies. All living organisms do this at the atomic and molecular levels, even the crudest microorganisms; the more advanced an organism is, the larger and more organized in themselves are the outside structures that it is able to use and transform into its synthetic body.

Artificial systems of investigation, control, and communication, as these are, have a name: technology. The body of a living organism is its technology; the technology of an organism is its body. Our life is, after all, only the pattern of information spelled out in our genes: a pattern that survives any given atom in our bodies, except for the ones we have not yet metabolized at our death. Our own technology is an extension of our bodies, but our bodies are nothing more than such cumulative extensions. Biological evolution, and arguably even prebiological evolution, are in this sense precisely the increase in the complexity and power of technology. Nature is technology, then, and if nature is the body of God we may formulate a surprising ninth axiom: *God is the process of increasing technology.*

If our moral function is to serve God, then it is to help God change from a fetus into a fully developed being, to realize God's future growth and self-awareness. The way to do this is to continue to innervate the universe by knowledge and control, thereby extending our own bodies, the region of our own technology, throughout the universe. Thus the tenth axiom: *To serve God is to increase the scope, power, beauty, and depth of technology.*

Our logic has brought us to an astonishing and perhaps shocking conclusion, utterly at odds with the prevailing mood of our culture. How can we redeem this statement, and make it fit what we feel about our role in the world?

The answer must involve a thorough reevaluation of what technology is and what we mean by the term. We know there is such a thing as bad technology, but the theological implications we have discovered require that we define good technology, because without good technology we cannot adequately serve God, if God is conceived of as being within nature. It will no longer be sufficient for us to attempt to get away from or to dissolve our technology; even if this were possible, it would be to deny our divine duty and commit a sin against the spirit of nature. Our investigation of what good technology is may have the virtue of clarifying what is bad technology, bad service of God, and thus may constitute a powerful if gentle critique of society.

Good technology, first of all, increases and does not decrease the organized complexity of the world. The science-fiction writer Ursula Le Guin has a lovely phrase for this: the purpose of the Ekumen, the loose confederation of intelligent species she describes in her novel *The Left Hand of Darkness*, is "to increase the intensity and complexity of the field of intelligent life." Perhaps there is a little biochauvinism in the formula—we might also want to encourage beautiful crystals and sculptures—but it still will do very well. Good technology respects the existing technology of nature, and even when adding to it does not destroy the complex order and beauty that helped it evolve and upon which it is based. Bad technology is technology that destroys technology, whether in the form of the bodies of animals and plants, or in the form of our own rich material and mental culture.

As a direct implication of the injunction to increase the organized complexity of the world, good technology preserves earlier stages and products of its own process. It will, therefore, pay special attention to the preservation of chemical complexity, to the preservation of the richness and variety of life, to the preservation of the higher organisms in particular, and to the care and reverence of human life. This is the natural order of our increasing concern, because life, higher organisms, and human beings are closer and closer approximations to the emerging nervous system of God. Likewise, within an organism we give preference to the higher functions, especially the nervous system, over lower vegetative functions. This hierarchy is really common sense; it is the automatic assumption of any good surgeon or any animal caught in a trap in making decisions about which part must be sacrificed to save the rest. Indeed, it will be necessary to replace certain environmentally unsound technologies with the more efficient, elegant, and benign ones that the new science is making possible. We will need to isolate fine examples of ancient and unique ecosystems in "wilderness areas"—however misleading the term may be—from the natural interference of other, highly competitive species, such as ourselves and the pantropic weeds, in order to promote the richer evolution of the rest. This is exactly what some environmental radicals have demanded, though on other grounds, but we do not have to yield to an antihuman and antitechnological ideology to make such choices.

The theology outlined here would suggest that we embrace an activist, restorationist environmentalism that goes with, not against, the natural inclination of humanity toward greater experience, self-awareness, mutual feedback, and technical power. Our job is not to leave nature alone or to coexist peacefully with it; we *are* it, we are its future, its promise, its purpose. Its future landscapes are partly up to us.

Creating the Landscapes of the Future: Innate Resources

The creation of landscape is a special kind of human activity. To do it well, and especially to teach it to others, a sound working theory of the human capacities and talents involved might be valuable. Are we even capable of such an enterprise?

We can begin to answer this question by asking another: what other human activities would ecological invention resemble? Science, certainly; but science is normally considered a method of analysis, and ecological creation is a process of synthesis, of performance. Ecological creation would indeed be guided by analysis; it would serve as a test of the analysis it relies on and be richly suggestive of new analytic models, new hypotheses. But this fertility of hypothesis is called into being by the imperative of successful and positive action. Unlike science, which professes to understand the world by external observation and measurement, ecological creation must understand the object of its study from the inside.

But is science as detached from the world it studies, as purely analytical, and as divorced from performance as its stereotype implies? In quantum physics what the physicist observes is partly the result of the process of observation. The accelerator now creates particles to be studied that are very rare or even normally absent on this planet. Chemists, with the enthusiasm of artists, regularly create molecules that never existed, and biochemists produce chimeras stranger than any generated by mutation or bacterial gene transfer or sexual recombination. What ecological creation is challenged to do is to create an ecology that is indistinguishable from a natural reality, though the first of its kind.

But if ecological creation is not exactly a science, neither is it only a technology, in the traditional sense of the word, for its goal is not a product but a process. One could say that the biological "machine" the creative ecologist produces must first attend to its own ordered reproduction. Like the natural technology of animals and plants, its work is itself. Technology cannot guide itself; our earlier, crucial distinction between bad technology and good technology needs criteria that cannot be found within technique alone. Nor is ecological creation quite like a craft, in which the boundaries are clearly established and the criteria of productivity can be easily measured. In some ways it is almost more like play, a serious kind of play that submits itself to high and demanding canons of perfection, and which is not embarked on only for its own sake.

In fact the activity that inventionist ecology most resembles is art. First we must know what we mean by "art." A simple definition may be best: art is the creation of beauty. But what is beauty? Here, oddly enough, the perspective of the biologist may be more instructive than that of the aesthetician. Recent work in the biological foundation of aesthetics (which I have summarized and explored in my books *Natural Classicism: Essays on Literature and Science*; *Rebirth of Value: Meditations on Beauty, Ecology, Religion, and Education*; and *Beauty: The Value of Values*) suggests that the aesthetic capacity is an inherited and sophisticated competence. It is given specific form by culture, as is our predisposition to speech, but its deep syntax, as it were, is wired into our neural hardware. Obviously, then, aesthetic perception is designed for some very important function, and was selected for and refined by evolutionary mechanisms; we share this capacity with other animals, but in us it is enormously more developed, flexible, and variable in application.

Recent research suggests that the aesthetic sense is a capacity to organize and recognize meaning in very large quantities of ill-defined information, to detect and create complex relationships and feedback systems, to take into account multiple contexts and frames of reference, and to perceive harmonies and regularities that add up to a deep unity—a unity that generates predictions of the future and can act as a sound basis for future action. The new understanding of the vigorous constructive activity of the sensory cortex suggests that we do not passively receive the world of vision, but actively create it (see David Marr, *Vision*). The aesthetic capacity is to perception as perception is to mere sensation, or as sensation is to the mechanical effect of some outside event upon an inanimate object. In other words, the unities that the sense of beauty appreciates are a higher, more integrated version of the enduring solid objects in space and time that the sensory cortex constructs for our use out of the storm of disparate information that hits our sensors. A complex unity like an ecosystem is perceived by the aesthetic sense as a solid object in space is perceived by sight. The exercise of the aesthetic capacity is rewarded and reinforced by the brain's own self-reward system, which we feel as aesthetic pleasure.

A new generation of scientists, including Ilya Prigogine, Mitchell Feigenbaum, and Benoit Mandelbrot, has recognized the importance in nature of dissipative, nonlinear feedback systems, in which causality is not one-way but rather mutually circulated among all of the system's components or turned recursively back upon itself. The popular term for this new science is "Chaos," though more accurately it deals with the order in chaos and the chaos that can arise out of deterministic order. Much of this work is summarized in James Gleick's important book *Chaos: Making a New Science*. "Chaotic" systems have the peculiar characteristics of unpredictability and self-organization, and seem to be drawn toward "strange attractors," outcomes whose graphic representation is immediately recognizable as beautiful. Chaotic attractors exhibit the characteristics of self-similarity (their patterns are repeated, often with variations, at an infinite variety of scales) and fractal discontinuity (they never fill space homogeneously with straight lines or flat planes or solid volumes, but instead are always frilled or braided or pitted or interpenetrated with empty space, though they can, paradoxically, approach infinitely close to filling the available space as they are plotted). Such forms denote the feedback processes—of which biological evolution is one—that created the universe as we know it. There is good reason to believe that those forms appear beautiful to us because we have inherited from our own evolution a capacity to recognize, and to participate in, the creative processes of nature. We find a healthy and ecologically rich landscape beautiful because it embodies, as living attractors, the forms that are generated by self-organizing feedback systems. Thus our innate aesthetic abilities are a reliable rule of thumb when we wish to judge the ecological viability of a natural landscape.

What capacity could be better adapted than our aesthetic sense to the complex, context-rich work of ecological invention, which must harmonize into a higher unity large masses of mutually dependent information? Our destiny as a species now appears to be bound up with the success of our attempts to reconstruct our living environment (not only on this planet, perhaps). The sense of beauty tells us what is relevant, what is likely, what is proper, what is fruitful. We would be in a desperate case if the only capacity we could rely on was our logical ability to put two and two together; there is not enough time to work everything out in that fashion, and there is simply too much information, and too many possible consequences, to do so without those higher integrative abilities. Aldo Leopold suggested the existence of a human natural aesthetic; ecological invention may now with some justification call upon a human aesthetic capacity that is not a merely passive appreciation and in which the artificial and the natural cannot be distinguished. Indeed, one of our problems is this distinction: true gardeners of the planet will no longer need it.

Creating the Landscapes of the Future: Historical Resources

Every major art form exists within a tradition, a context of prior practice that serves as the inspiration and raw material of the artist's conceptions. On what tradition can the environmental artist call?

We have at our disposal—though we have used it very little—a rich storehouse of theory and recorded experience in the field of art and nature: Renaissance aesthetics. The Renaissance already knew, for instance, that the simple recursive algorithm or feedback process that generates the Fibonacci series (add the two previous terms in the series to get the next) could be used both to predict the demographic future of a naturally reproducing population of animals, and to generate what might be called the simplest of all fractal curves, the beautiful Fibonacci spiral. That spiral can be found in seashells and sunflower heads and throughout nature in general. The Renaissance also knew that the ratio between any two successive terms in the series could provide, with increasing accuracy as the series was extended, the Golden Section ratio, which was the touchstone in architectural and artistic proportion of that time.

The Renaissance poet Sir Philip Sidney, author of the *Arcadia*, defines art as the imitation of nature. He does not mean that the artist imitates what nature merely *is*, as a photograph or a diorama copies the visual externals of a scene, but rather imitates what nature *does*, that is, generate a living and self-developing order. Human art, he maintains, can better the current productions of nature, but precisely because human art is a natural process. Shakespeare says the same in *The Winter's Tale*. The shepherdess Perdita has just declared that she won't have carnations or "streak'd gillyvors" in her garden because, like some American environmentalists, she disapproves of the fact that they have been bred and hybridized by genetic technology.

PERDITA: Of that kind
Our rustic garden's barren; and I care not
To get slips of them.
          . . . For I have heard it said
There is an art, which in their piedness shares
With great creating Nature.

POLIXENES: Say there be;
Yet Nature is made better by no mean
But Nature makes that mean; so, o'ver that art,
Which you say adds to Nature, is an art
That Nature makes. You see, sweet maid, we marry
A gentler scion to the wildest stock,
And make conceive a bark of baser kind
by bud of nobler race. This is an art
Which does mend Nature, change it rather; but
The art itself is Nature.

                                        (4.4.83-97)

We find similar ideas in other great Renaissance aesthetic theorists—
the architects Alberti and Palladio, the critic Scaliger, the philosophers
Pico, Ficino, Vida, the scientists Bruno, Hariot, and Bacon. Those ideas
were the guiding principles of Leonardo da Vinci. In the Enlightenment
they informed the great landscape gardeners of England, like Capability
Brown. They appear too in Marvell's "Upon Appleton House," Pope's
"Twickenham Garden," and in the marvelous scene where Elizabeth
Bennet visits Darcy's country house in Austen's *Pride and Prejudice*.

Renaissance art gave rise to good science; its effort to imitate nature had
profound heuristic value. The famous discoveries of Leonardo da Vinci in
anatomy, aerodynamics, hydrostatics, and mechanics grew out of his attempt to imitate nature in drawing, painting, sculpture, and architecture.
Brunelleschi, in his endeavor to imitate the natural phenomenon of visual
perspective (whose discovery he—perhaps erroneously—attributed to the
ancient Romans), invented pictorial perspective, and in so doing made large
advances in the science of projective geometry. The great Elizabethan scientist Thomas Hariot, a close friend of the poets Christopher Marlowe, Edmund Spenser, Walter Raleigh, and Philip Sidney, and of the painter Nicho-
las Hilliard, made major discoveries in optics, algebra, biology, and New
World anthropology, largely because his friends and patrons demanded of

him practical principles for the reconstruction of nature. (I discuss these matters in the chapter on the School of Night in *Natural Classicism*. If good science comes out of the artistic imitation of nature, we have a warrant that such art is itself an extension of the work of nature. The Elizabethans thought that we know by doing, and the best and highest thing we can do is imitate the creative activity of nature.

Our assessment of the human resources that can be deployed for the work of environmental invention presents us with some curious but suggestive conclusions. First, we must look to the artistic and aesthetic capacity, and learn more about its roots in our biology and evolution; the peculiar kind of art and aesthetics involved are those of the imitation of nature. Next, we may call on Renaissance aesthetics for the nucleus of a sound body of theory about the artistic imitation of nature. We can extend that body of theory by the study of chaotic self-organizing feedback systems. Finally, we may find in the practice of art the heuristic value of imitation: the attempt to reproduce accurately the functions of nature forces the artist not only to increasingly close observation, but beyond, to increasingly stringent experimental tests of ideas. This labor is not merely analytical, but creative, and its natural reward is beauty.

Arcadias, Terrestrial and Extraterrestrial

If the artistic/scientific imitation of nature to create new landscapes were a new project, we might have cause to be anxious, on the grounds that though our theory might be sound, the stakes—the health of the planet—are too high. Two arguments address this reasonable objection.

The first is that human beings have in fact been creating new ecosystems for thousands of years. The arcadian tradition of the farmed *rus* or countryside (*rus* is the Latin root of "rural" and "rustic") is an ancient library of techniques by which humans can live harmoniously in a landscape they have partly created. Genetically tailored species such as wheat, grapes, cattle, and fruit trees are interspersed with husbanded forests, wild game, and carefully managed rivers and streams. Examples of this arcadian landscape exist all over the world—in the wine countries of Tuscany and Provence, in the rice paddies of Java and Bali, in the English Cotswolds, and in the ancient farms of China.

This folk art has risen at various times and places into the status of a high art, with conscious canons and some abandonment of purely utilitarian criteria of success: the art of the garden. We can trace its tradition in the history of Chinese, Korean, and Japanese gardens, and in the West we note the great lineage that runs from the hanging gardens of Babylon, through the Greek garden as described by Homer in his mythical isle of Phaiakia, the Roman gardens of Virgil, Horace, and Cicero, on into the Renaissance gardens of Italy and France, the English neoclassical gardens (Stourhead, for instance), and the gardens planted by Capability Brown and Gertrude Jekyll, to the impressionist gardens, Monet's Clos Normand at Giverny and Vita Sackville-West's Sissinghurst in Kent, and across the Atlantic to the American gardens of Longwood and Dumbarton Oaks. These gardens are part of the aesthetic education of the creator of ecosystems; they are epitomes of the arcadian landscape. If we can sensibly combine our new scientific knowledge of ecology and genetics with the old empirical wisdom of the peasants and farmers and gardeners, and the influence of our existing garden aesthetics, there is no reason why the Earth should not under human care end up with an even greater richness and variety of ecosystems than it now possesses.

The other reason why we should not be too anxious about the idea of the created landscape is that we will not stay on this planet forever. We must conserve and preserve the life of the Earth, but there are also dead planets out there that might be brought to life without risk. NASA is already seriously researching the proposition.

Of course, we need to know much more about how ecologies work before we take any irreversible steps in this direction. We particularly need a better bacteriology and a better understanding of the subtle interplay of plant, animal, and human societies, gene pools, and the climatological and geological feedback loops. Evolutionists and ecologists, who sometimes do not seem to talk to each other, need to come together for a grand synthesis. The best way to do this is through the practical craft of ecological restoration. We learn how ecologies work by recreating them, and in the process of recreating ecologies we will create a community of scientists and restorationists who will devise the arts of seeding life on other planets.

We also need to know much more about genetic inheritance and genetic expression. It is beginning to look as if the 95 percent of the genome that is not expressed is actually a jumbled but fairly complete archive of a given organism's entire evolutionary history. As with certain big old business computer programs, which have been patched and augmented so many times that the programmers no longer know what might still be useful, it is simply too expensive to clean out all the old material and really very inexpensive to store it in a dormant state. Further, the bacteria and viruses of the world constitute a huge lending library of past genetic diversity from all other living species. Using recombinant DNA techniques (as bacteria do all the time) it may be possible one day to reconstruct and resurrect extinct species from this "fossil" DNA. We may thus eventually be able to undo the damage we have already done to species diversity, and perhaps even to restore whole ecosystems that existed before the advent of humankind.

Eventually we may modify existing species by gene tailoring, and even develop new species adapted to new ecological niches. Sooner or later we will leave the confines of this planet. When we do we may carry with us the seeds of earthly life, hardened and redesigned to thrive in alien environments and perhaps to transform those environments, as primitive life did on this planet, into a habitat for other more advanced earthly lifeforms. In this work we may become the seed-vectors and pollinators of the universe, carrying life beyond the fragile eggshell of this planet, so exposed to sterilization by a stray asteroid strike or an extra-large comet. We will eventually be in the business of the ecotransformation of planets—in fact we already are, with this one. We need to start thinking in these terms, and I have called for a commitment by our civilization to an eventual transformation of the dead planet Mars into a living ecosystem. We should do this not only because it is a noble thing to do in itself, but also because we will not ever know with any confidence how our own planetary ecosystem works until we ourselves have created one on another planet.

The Poetry of Ecogenesis

I chose this work of planetary transformation as the subject of my epic poem *Genesis*. The action of the poem covers the major historical events of the period approximately from 2015 to 2070. A group of scientists and technologists, led by Chancellor ("Chance") Van Riebeck, is charged by the United Nations with the scientific survey of the planet Mars. Using theories derived from the Gaia hypothesis, they clandestinely introduce hardy genetically tailored bacteria into the martian environment with the intention of transforming the planet so that it would be habitable by human beings. The Earth has fallen under the theocratic rule of the Ecotheist Movement, which divides human beings from the rest of nature and regards all human interference with nature as evil. Chance and his followers are captured and put on trial, and war breaks out between the martian colonists and the home planet. Though Chance and others lose their lives, the colonists are able to gain their independence by threatening to drop a moonlet on Earth. After a bitter renewed struggle led by the hero Tripitaka the colonists obtain a complete inventory of earthly life-forms, sometimes called the Ark, or the Lima Codex. With the help of this inventory, and led by Beatrice Van Riebeck, they complete the terraforming of the planet. A religious leader, the Sibyl, is born to the colonists; her teaching reconciles the ancient mystical wisdom of the Earth with the new science and cultural experience of Mars.

The scientific and technological material of the poem constitutes not only a large part of its content but also a gigantic metaphor of its structure and form. In other words, the unwritten poem is the barren planet, and the composition of the poem is its cultivation by living organisms. But the word "metaphor" fails to capture the dimensions of the trope that is the poem, for the gardening of Mars by the code (or "codex") of life is act, theme, myth, argument, and form at once. The forking tree of evolutionary descent is the forking tree of grammatical and logical construction, the forking tree of plot and story, the forking tree of aesthetic form, the forking tree of family descent, and the forking tree of human moral decision. Those trees are in turn connected as branches to the stem of the great tree of the universe itself.

I conclude this essay with a couple of passages describing the gardening of Mars by its master gardener Beatrice Van Riebeck.

> It is a matter very practical:
> The gardening of crater planetscapes.
> Few books record its arts and its techniques;
> Yet Cicero's landscape gardeners would know,
> When they laid out his grounds by Lake Lucrino,
> And the patricians of the Alban Hills,
> Who set their villas by the crater-lakes
> Of Nemi and Albano clad in vines
> And let their grottos give a prospect on
> A glimmering water, framed in shady pines—
> They'd be worth asking, if she might invoke
> Their gentle, haughty shades for such discourse;
> Yet they passed on their wisdom, as the Greeks
> Did to the Romans and the Romans to

The masters of the Renaissance; they taught
The gardeners of England how to shape
A sylvan walk to imitate the trials
Of Hercules or sharp Odysseus,
Instruct a guest-Aeneas how to choose
The way of piety and fortitude.
And they, in turn, taught the Americans:
The gardens of Dumbarton Oaks, and those
The Du Ponts planted outside Wilmington
Carried the same hermetic wisdom on
Across the oceans, and the garden-worlds
That glitter in a necklace round the sun
Bear the same history, the land of shades
Transformed to paradise, to fairyland,
To purify the dreaming of the tribe.
It seems that Beatrice must write the book,
Though, and reveal its secret name as Mars . . .

Sing then, sweet bride-ghost, mother-mariner,
Of the garden planted in the vales of Mars;
How Beatrice bled the themes into each other
Of native waywardness and Arcady.
First, though, let us recall how it had been
Before the helmed conquistadors had come;
What lay below the gold wings of the ship
That bore the speechless astronauts to ground.

A numb plain spread with stones. A weary steppe
All bleached to tired red with ultraviolet.
Soil crusted, sere; limonite, siderite.
Hard radiation in a waste of cold.
Rocks sucked dry by the near vacuum.
Stunned with the blank math of the albedo
The eye tries to make order of it, fails.
Whatever's here has fallen from someplace else.
Sometimes a crag a foot high, or a mile;
Always the sagging tables of the craters,

The precise record of a mere collision.
And yet a stunted and abortive chemistry,
A backward travesty of life, proceeds:
Parched cirrus cloud move over the ejecta;
A hoarfrost forms upon the shadow sides;
Dark patches colonize the regolith;
Sometimes with a thin violence a sandstorm
Briefly makes shrieks of sound between the stones;
Rasps off their waists and edges, and falls silent.
Time here is cheap. A billion years can pass
Almost without a marker; if you bought
A century of Marstime in the scrip
And currency of Earth, you'd pay an hour
Or a half an hour of cashable event.
It's really a young planet then, a bald
And mild mongoloid, a poor old cretin
Worth but a handful of Earth's golden summers.

And it was beautiful. Those who first walked there
Said it was fresh as the true feel of death,
As Kyoto earthen teaware, as the Outback;
As clean as is geometry, as bones.
To spoil this archetype, this innocence,
Was to incur a guilt whose only ease
Was beauty overwhelming to the loss,
Was a millennial drunkenness of life
That might forget its crimes in ecstasy.
But it was not enough to reenact
The long sensualities of mother Earth,
To take on trust the roots of history;
They must be minatory, and exact
A last accounting of the failed balance
All the intestate dowager had left.
There must be new assumptions in the matrix,
New ratios, dimensions, and arrays;
Beatrice finds her trope in simple mass,

The crazy lightness of all things here
Set, in a poetry that brewed delight,
Against the literal dimness of the light.

After the riot, then, of Earth diseases,
After mycosis, planetfall, the plague
And infestation of the weeds, the jungles
Of a lifeforce as fresh as it was vulgar,
The time came to prune and shape the flow.
On Mars all these fall slowly, dreamily:
Waterfalls, billowy, like the clawed waves
In Hiroshige prints of sudden storms.
Snow, in soft bales or volumes, scarcely more
Than bright concatenations of a vapor.
Rain, in fine drizzles, dropping by a cliff
Stained by the rocksprings and the clinging mosses.
Rivers and streams, whose wayward pressures thrust
More, by inertia, at their banks than beds,
And so can spread in braided flats and strands
To glittery sallow-marshes, quiet fens.
Ocean waves, swashy, horned, and globular,
Like the wave-scenery of an antique play
(Bright blue horned friezes worked to and fro,
A fat-lipped leviathan, and a ship).
And then on Mars all these rise swifter, easier:
Smoke, which makes mushrooms in the wildest air,
Fountains, which tower and tower, whose very fall
Is caught up once again within the column
Of their slow and weighty rise; yes, fountains
Shall be the glory of our Martian gardens.
Flames, in like fashion, scarcely dance on Mars
So much as dart into the air, like spirits
Lately penned in earth but now set free.
And the warmed thermal plumes from open fields
Of ripened grass or stubble here make clouds
As tall as chefs' hats, stovepipes full of thunder.

The poets of Earth refine the fuel—
The hot benzene of value culture burns
To power its subtle engines of desire—
From fossil liquors buried in the stone
Through ages of creation and decay.
They can afford to toss aside the raw,
And take for granted a world cooked and rich
With ancient custom, languages numberless
As layers of autumn leaves within the forest,
Nature itself grown conscious, turned upon
Itself to make its rings so intricate,
History fertile with its own grave-mould.
And so the poets' work is little more
Than cracking out the spirit they inherit
In the tall silver towers of poetry
To brew those essences, those volatiles,
Those aromatic esters, metaphor,
Image, trope, and fugitive allusion.
Prodigals, they burn half that they use
To purify the rest; and they make little,
Only a froth or lace of ornament.
The poets of Mars must brew the very stuff
The Earth poets burn as waste; must mate each word
With breeder's care, and dust the yellow pollen
Over the chosen stamen; graft the stem
To coarser stock, and train the line to sprout
Productive variation years ahead.
The poets of Mars must make the myths from scratch,
Invent the tunes, the jokes, the references;
They must be athletes of the dream, masters
Of the technology of inventive sleep,
Architects of the essential shades of mood.
What they inherit from the Earth, they earn,
Through sacrifice and trouble, and they breed
More than they are bequeathed. So Beatrice,

Taking into her hands her garden tools—
A dream of a Campanian burial,
A trope of lightness, and a wild new world—
Begins the cultivation of the void.

This garden: let it propagate itself,
Sustain itself, an arch-economy
Dynamically balanced by the pull
Of matched antagonists, controlled and led
By a fine dance of feedbacks, asymptotic
Cyclical, damping, even catastrophic.
Let there be forest fires to purge the ridges;
Let there be herbivores to mow the parkland,
And predators to cull their gene pools clean,
And viruses to kill the carnivores
That sheep may safely graze. Each form of life
Shall feed upon the wastes of its convivors;
Let there be beetles and bacteria
And moulds and saprophytes to spin the wheel
Of nitrogen, corals and shells to turn
The great ratcheted cycle of the carbons;
Each biome—grassland, forest, littoral;
Benthic, pelagic; arctic, desert, alp—
Shall keep appointed bounds and yet be free.
Let the new species bud and multiply;
Let monsters speciate and radiate
And seize the niche that they themselves create;
Let some be smothered or extinguished; some,
Effete, exquisite—the trumpeter swan,
The rare orchid, the monoclone cheetah—
Cling to some microclimate or kind vale,
Eking survival for a clutch of genes.

Beatrice has a wand, a metatron,
To help her in her work; a golden bough
Wherewith she will transmute and charge her world
With metamorphosis: the flowering plants.

Mars was first seeded with the gymnosperms:
Horsetails, treeferns, cycads, conifers.
Now comes the carnival of angiosperms,
The spirits of a world made up anew
In all the colors that vibrate the field
Of time's ether, giving a taste to light.
As each seedclone package comes from the shops
Of sleepless Charlie and Ganesh, she breaks
It out among her helpers: Hilly Sharon,
Her pilot, flies her among the green coombs
And verdant-headed hills and plains of Mars
To supervise the setting of the seeds.
Ganesh has multiplied the speed of growth
Tenfold, relying on the lesser weight
The plant must carry to its destiny;
And soon that glorious, pathetic fate
Bursts in a thousand blooms across the planet
In petals bluish, pink and mauve and gold.
Now the first bees, their wise, conformist brains
Not muddled much by the new dispensation,
Blunder among the anther and the pistil.
Quivering beech-groves, the work of but a year,
Rootle and creak on hillside or in canyon,
Their white feet sunk in a sweet red mould.
From Earth we saw it through our telescopes:
Whole hemispheres turned white with fragrant daisies;
Reefs rising in long chains and rings about
Those windy coasts under a tiny sun.

      (*Genesis* 4.4.282-312; 4.5.53-237)

REFERENCES

Botkin, Daniel. 1990. *Discordant Harmonies: A New Ecology for the Twenty-First Century*. New York: Oxford University Press.

Gleick, James. 1988. *Chaos: Making a New Science*. New York: Viking.

Le Guin, Ursula. 1983. *The Left Hand of Darkness*. New York: Ace Books.

Marr, David. 1982. *Vision*. New York: Freeman.

Turner, Frederick. 1988. *Genesis, an Epic Poem*. Dallas: Saybrook Publishing Company.

_____. 1991. *Rebirth of Value: Meditations on Beauty, Ecology, Religion, and Education.* Albany: State University of New York Press.

_____. 1992. *Natural Classicism: Essays on Literature and Science.* Charlottesville: University Press of Virginia.

_____. 1992. *Beauty: The Value of Values.* Charlottesville: University Press of Virginia.

PART II

THEORY

Restoration or Preservation? Reflections on a Clash of Environmental Philosophies

G. Stanley Kane

Ours is an era in which there is little left of nature that has not been extensively altered by the activities of human beings. Among proposed remedies are preservation, setting aside areas that still remain undisturbed and protecting them against human encroachment, and restoration, bringing degraded areas back to an unspoiled condition. On first thought one might suppose that preservationists and restorationists would make natural allies, but even a cursory reading of the relevant literature shows that all is not harmony and peace between the two groups. The writings of William Jordan and Frederick Turner, for example, include some surprisingly sharp criticisms of preservation. In this essay I wish to assess the dispute between the restorationists and preservationists, centering my discussion around two basic philosophical issues: the concept of nature and the relation of humankind to nature, and the character of human knowledge.

Throughout this essay I use "restoration" to refer to the work explained and interpreted under that heading by Jordan and Turner. There are other restoration projects, and doubtless other ways of interpreting restoration, but these fall outside my purview.

The Restorationist Critique of Preservation

The nub of the critique of preservation is the claim that it rests on an unhealthy dualism that conceives nature and humankind as radically distinct and opposed to each other. Jordan and Turner offer little evidence to support this indictment, but others have, sometimes pointing to the Wilderness Act of 1964 as especially telling (Callicott 1991, 240). According to this act, a wilderness is an area where "in contrast to those areas where man and his works dominate the landscape . . . the earth and its community of life are untrammeled by man, where man himself is a visitor and does not remain. . . . [It is] an area . . . retaining its primeval character and influence, without permanent improvement of human habitation, which is protected and managed to preserve its natural conditions and which generally appears to have been affected by the forces of nature, with the imprint of man's work substantially unnoticeable . . . " (Devall and Sessions 1985, 114–15).

Dissatisfaction with dualism has for some time figured prominently in the unhappiness of environmentalists with mainstream industrial society (Merchant 1980; Roszak 1973; Berman 1981). Jordan and Turner turn the critique of dualism against preservation-oriented environmentalists themselves. In their view preservationists are imbued with the same basic mind-set as the industrial mainstream, the only difference being that the latter exalts humans over nature while the former elevate nature over humans. According to the restorationists, neither position is healthy. One underwrites exploitation, with devastating environmental consequences; the other effectively takes human beings out of nature altogether and makes wilderness of it (Turner 1985, 48).

In the judgment of the restorationists, the exclusion of humans from nature deforms both. Set off against nature, humans can only work harm in the world. Any possibility of constructive stewardship is denied them, and the best they can do for nature is depart it and leave it alone (see Turner, "The Invented Landscape," this volume). But nature suffers as well in this separation from human beings, because it is deprived of the services that humans render as rightful citizens of the biotic community. Dramatic testimony to this is seen in Turner's statement that wilderness areas from which humans are systematically excluded are "the most astonishingly unnatural places on earth" (1985, 45).

What are we to make of this criticism of environmental preservation? In answering this question we need to distinguish the issue of the merits of dualism as a philosophical outlook from the question of whether preservationists are really dualists. I am persuaded that many of the faults found with dualism by its detractors not only are real but have been fateful. But is the preservation program really committed to these errors? There is good reason, I believe, for thinking not. We can see this if we look in two places: first, at the complete environmental program supported by most preservationists; and second, at the logic of preservation itself.

It might make sense to ascribe the nature-humanity dualism to preservationists if wilderness preservation were the whole of their environmental program. It would make even more sense if in addition their principal reason for seeking wilderness preservation were the conviction that nature can be fully itself and thus have full value only when left undisturbed by human beings. Though there are exceptions, preservationists typically do more than just sponsor wilderness preservation. They also work actively on a broad array of environmental issues, such as air and water pollution, toxic waste, soil erosion, global warming, and so on. To think that such preservationists are fundamentally inspired by the nature-humanity dualism and a misanthropic view of human beings is not at all a necessary, or even a very reasonable, inference. To be sure, they *are* worried about the impact that humans are now having on natural systems, and they do think that human activity at the present time is alarmingly destructive of nature. But so do many others, including restorationists, who would not think of solving the problem through a policy of apartheid for humans and nature. It makes more sense to think that these preservationists are driven, not by the notion that human contact and commerce with nature should be kept to a minimum, but by the desire that humans avoid the kind and the magnitude of interaction with nature that destroys the health of the world and the beings, human and nonhuman, to which it is home.

The definition of wilderness in the Wilderness Act, as land kept free from the influence of human beings, might seem to count against this conclusion. This is a definition of wilderness, however, not of nature in general; that a person support the protection of wilderness and still recognize a legitimate need for other types of land devoted to other purposes is wholly consistent with this definition. It is hard to see how someone willing to accept multiple land uses in this way is a victim of nature-humanity dualism.

In any event, there is no logic that requires dualism as a philosophical underpinning for preservation. Dualism might support preservation, but it is not the only outlook that would do so. Preservation could be grounded just as securely on the much more innocuous premise that there are limits to the freedom of human beings to use nature solely for their own purposes.

There seems, then, no compelling reason for thinking either that dualism is implicit in preservation or that its practitioners generally think it is. It is perhaps puzzling that Jordan and Turner do not see this, but more puzzling, I think, is the sharpness and relentlessness of their attack on the preservationists, accentuated by the fact that they offer little, if any, criticism of those who have plundered the natural world and left it standing so desperately in need of the healing powers of their own art of restoration. They pay no attention to the obvious point that in our present situation still-untouched lands not accorded legal protection will sooner or later almost certainly suffer the fate that has historically overtaken virtually all untouched lands in the path of industrial progress. The value of preservation as means of limiting further ecological destruction is not once acknowledged. We see here a curious phenomenon: a movement that desires to restore the earth to a more natural condition singles out, from all the parties active in public life today, a group that wishes to preserve some lands in their natural condition, and belabors them for unhealthy attitudes toward nature.

Restorationism on Nature and Humanity

The first principle of restorationism is that nature and humanity are fundamentally united rather than separate. Humans are a natural part of nature. The familiar distinctions of the natural and the artificial, of nature and culture, of ecology and economy, are not oppositions but a series of diverse and interrelated elements within a rich and unified whole. Human life, in all its manifestations, depends on nature and is an outworking of the same forces that are at work throughout the biosphere, indeed throughout the universe. But equally, because humans are an integral part of the natural order, nature also depends on humanity, and cannot maintain full health and integrity without the activities of human beings. Nature and humanity are thus interdependent, and as a consequence their proper relation is cooperative, not adversarial. When each carries out its own proper functions, they work together to produce results that are wholesome and beneficial for both.

Compared with the dualism considered earlier, this scheme of understanding allots a much more positive role to humans in their interactions with nature. No longer are we either excluded from the world or condemned to exploit it. Instead, human participation is essential both for our own good and for that of the world. We can now feel at home in the world, full-fledged citizens of the land community, beings who belong where we are, in a place that requires of us the vital work of stewardship—a critical form of which is restoration.

In this outlook human use of nature is not something to be decried. Humans have a wide range of legitimate needs, all of them bred into us by nature itself. Our instincts and capacities for satisfying these needs, including the astonishing intellectual and technical abilities our species has acquired over the centuries, are also products of nature. Nothing is intrinsically amiss or unnatural when we use these abilities in our dealings with nonhuman nature. To be sure, we often bring about substantial change in the nonhuman world, but that is fully natural and not to be deplored. Nature is a dynamic realm, a domain of incessant change sustained by the actions and reactions of its constituent parts. This does not mean that humans cannot do serious ecological damage or that we should not try to prevent or control such damage, but it does mean that humans need feel no hesitation about manipulating nature solely on the grounds that it leaves its mark on the world.

When it comes to assessing this restorationist outlook, the most serious issue, in my judgment, does not concern the merits of this conception of the relation of nature and humanity. Rightly understood, I believe this conception is markedly superior to that in dualism. The crucial question about the restorationist outlook has to do instead with the degree to which the restorationist program is itself faithful to its own vision of the relation of humans to nature. Rejecting the old domination model, which sees humans as over nature, endowed with authority to dominate and control it, restoration theory champions a model of community participation. Yet some of the descriptions that Jordan and Turner give of what restorationists are actually up to in the overall economy of nature do not cohere well with the community participation model.

For example, Jordan thinks that "the fate and well-being of the biosphere depend ultimately on us and our relationship with it" (see Jordan, " 'Sunflower Forest': Ecological Restoration as the Basis for a New Environmental Paradigm," this volume). These words might mean only that we should discontinue or scale back the activities that threaten the biosphere, but for restorationists they signify considerably more. Turner explicitly states that it is time for us to renounce what he calls false ecological modesty, recognize that we are "the lords of creation," and "take responsibility for nature" (1985, 51)—a responsibility, he thinks, that extends to creating "man-made nature" (1988, 50). Restoration is part of this project of creating man-made nature. (Some might think that the capacity of human beings to damage the ecosystems of the earth automatically gives them a controlling role in the biosphere. But this is a mistake. That humans can harm the biosphere no more gives them special authority over it than the fact that I can injure or kill my neighbors makes me lord of the neighborhood. Special authority depends on something other than mere ability to destroy.)

It is hard to square the description of humans as the lords of creation with the community model of the relation of humans to nature. Indeed, Turner's comments seem to fit better into the domination model. Lords of the world, exercising responsibility for the fate and well-being of the biosphere, even to the point of creating man-made ecosystems, and beings who thus hold literal life-and-death power over the nonhuman realm, surely occupy a position of dominance, and everything else holds a place of subservience. Fellow members of a community, in contrast, are on more equal footing; they enjoy more independence and autonomy than any of the nonhuman participants in the lords-of-creation scenario. A lord of creation is not just one of many members in a community.

Another holistic model might be more serviceable to the restorationists, namely that of nature as an organism. As with the community model, this pictures nature as a system of interconnected parts. A fundamental difference, however, is that in an organism the parts are wholly subservient to the life of the organism, whereas members of a community have lives of their own apart from their functions in the community, thus possessing a measure of independence that parts of organisms do not have. The major parts of an organism are its organs; the members of a community are not organs, but are themselves organisms. They stand to the community, therefore, in a relation very different from that of organs to organisms (Katz 1985).

If we could think of the biosphere as a single living organism and could identify humans with the brain (or the DNA), or control center, of the biosphere, we would have a model that fits the restorationists' view of the role of humans in nature much more closely than does the community model. But just how plausible is such a model? Is there any credible evidence that humans are indeed the control center of the biosphere, or any compelling reason for thinking that they have the ability to carry out this function well?

The evidence is, to put it mildly, not strong. If we were the biosphere's control center, then the extinction of the human race would mean the death of the biosphere. Mass extinctions of the past have not had such a catastrophic consequence. It is difficult to see why the biosphere could not just as easily withstand similar extinctions now, even if humans were included in the die-off. What makes the human species so much more vital to the living earth than other species? According to one prominent analyst, the biosphere would be able to withstand even nuclear war followed by nuclear winter (Lovelock 1988, 232).

But let us suppose, contrary to the evidence, that humans really are the control center of the biosphere and that we really do have responsibility for directing its progress and ensuring its well-being. Why should we believe that we are up to the task? The prospects are not encouraging. The difficulties that would have to be overcome are just too great. For example, how could we expect to acquire the knowledge that would be required? The biosphere is too large and complex, too much the product of a long and intricate history of natural development, and too many of its processes are marked by an intrinsic indeterminacy, for us to comprehend it thoroughly enough to control its fate.

Even if we could acquire the necessary knowledge, we would have to face the problem of how to find the wisdom and moral will to employ that knowledge to beneficial rather than harmful purposes. The record of history inspires little confidence that humans in positions of responsibility will use their power generally to improve the health and well-being of those under their authority. Granted, some of the harm done has been a consequence of ignorance, and is thus correctable by advances in knowledge, but much of the damage takes place when people in authority use their positions to further their own interests at the expense of those in their charge. If that state of affairs is not remedied, gains in knowledge and power will only increase the potential for social and ecological disaster; for the more powerful our technologies are, the less tolerant they are of human error or ill intentions.

To consider humans as the control center of the living earth (or as lords of creation) is to ascribe to them a dominating role in nature. Is this significantly different from the old-fashioned domination model? If not, then restoration, notwithstanding its genuine concern for ecological wholeness and well-being, may be unable to make good on its promise to bring healing to nature.

Striking parallels exist between the old domination program and restoration. The most basic is that in both systems humans hold the place of highest authority and power within the world. Also, neither view recognizes any limits to the scope or range of legitimate human manipulation in the world. Everything is fair game for our manipulations if useful to our work. This does not mean that there are no constraints—only beneficial manipulations should be undertaken—but it does mean that nothing is intrinsically off-limits. A further parallel is that because the fate of the world rests on humans, they must have a clear idea of what needs to be done. They must know what conditions are good (or at least what conditions are better) and then work to bring them about. Their activity, then, requires them to shape the world after ideas in their own mind.

There are also important differences between the two theories. First, restorationists no longer view the world in the old dominationist way as a passive and inert object; instead it is a system that is alive, dynamic, creative, and one that has a history. Second, restorationists consider humans to be continuous with nature rather than separate from it. Third, though both assign to humans a controlling role in the world, dominationists conceive this in terms of conquest while restorationists conceive it in terms of healing. In restorationist doctrine humans are physicians to the biosphere, who through their special knowledge and skill aid nature as it drives to maintain and develop itself. A fourth difference is that restorationists recognize that needs and interests other than exclusively human ones have a claim upon us. Fifth, the ideas that must serve to guide our work in the world are drawn not solely from a consideration of human needs and purposes, but from an understanding of the biosphere—the beings, the systems and the interconnections that make it up, and the values embedded therein. Sixth, in the restorationist program, humans' controlling role in nature is not to be used solely for their own good, as dominationists thought, but for the health and well-being of the biosphere. Finally, because restorationists believe that the biosphere has needs and interests beyond narrowly human ones, they are more conscious than dominationists of our capacity to harm nature.

These differences are significant, but the continuing parallels raise troubling doubts about whether restoration is sufficiently removed from domination. The degree of anthropocentrism and human control that still remains in restoration, although not as crude as in domination, could still make restoration seem more a part of the environmental problem than its solution.

If the community model is best, we are seriously misguided if we act as the lords of creation, believing that if we don't make things happen for the well-being of the biosphere then the job won't get done. Community members hold responsibility jointly for the health and integrity of the community, and community values are not enhanced by one member taking on what others are better suited for. In the community of nature, nonhuman entities have their own stakes in the well-being of the biosphere and their own contributions to make in furthering it. The good of the biosphere requires that they be given the freedom to play their special parts. If they are deprived of opportunity to do this, the community suffers.

In situations of ecological breakdown, it is tempting for humans to think that they can save the biosphere by seizing control and restoring things back to order and health. But if the community model is correct, yielding to this temptation is self-defeating. There are constructive possibilities for the use of power, but only within limits. Beyond these limits a member simply has to trust the system—the processes and arrangements by which the community lives and on which it depends—to be sufficiently self-regulating, self-adjusting, and self-maintaining to survive the challenges and assaults that come its way. If the system becomes too impaired to survive, power grabs on the part of individual members will not save it. (I hope it is clear that, in speaking about trusting the system, I am not saying anything that would encourage submission to repressive political regimes. I am speaking here of community, not political systems inimical to community.)

Trust of this type is, of course, risky. It puts one at the mercy of forces beyond one's control, which may explain the appeal of the domination model. But life is inherently risky. A system in which there are no risks is a system in which there is no life. Beyond a certain point, the effort to control and eliminate risk does more harm than good, and taken too far becomes deadly. Humans are incontrovertibly creatures of forces they do not control. If the forces that brought forth life in the first place cannot be trusted to maintain it—provided they are given the leeway to do so—there is little basis for thinking that anything humans can do will save the situation. (Trust need not, indeed should not, be naive or uncritical.)

The argument for trust, however, is not just that we may as well bow to the inevitable. In a community trust is a positive good—even when relations are not going smoothly. We need only consider family life to see that this is so. When conflicts arise within a family, maintaining trust and respect for individual autonomy, though risky, is not thereby made less valuable or less necessary to family survival and prosperity. If community is the best model for our basic relations in the world, the quality of human life, and of life generally, will ultimately depend more on trust than on control.

From the point of view of the champions of community, the unwillingness of restorationists to make room for an ethic of trust in our dealings with nature, and their reliance instead on a program of control, is precisely what makes them fellow travelers with the old-fashioned dominationists. Like the dominationists they give humans a larger role than they are suited for, one for which they have neither the knowledge nor the moral wisdom to carry out well. They do, of course, propose to redirect human control from the final end of the dominationists, that of human empire over nature, to that of the health of the biosphere, but they do nothing to scale back the role of humans in the world, and nothing to correct the mismatch between their unlimited task and their limited qualifications. As a consequence, restoration offers the world no realistic protection against continuing social and ecological disasters of our own making.

The restorationist might respond that this criticism underestimates the significance of the shift in ultimate end. The crucial problem with domination, the response might go, is that it had no interest at all in the health of the biosphere. It is not surprising, then, that when Western civilization mobilized national economies to promote human empire, ecological values took a beating. Restorationists, however, recognize the supreme importance of ecological values. If the global economy were now to become reorganized around the principles and values of the restorationist program, humans could be expected to do much better in avoiding further ecological destruction in restoring the health of the biosphere.

There is some validity in this response, but it does not meet the basic force of the objection. It is on solid ground in holding that the chances of hitting a target are better when we are aiming at it than when we are not. But the dispute is whether we can act without mischievous consequences, no matter how unexceptionable the ends we seek, if we do not know very fully what we are doing. To the critic, the limitations of our knowledge are emblematic of our holding a more humble position in the biosphere than that of its lords. If we step out of our proper role and presume to take responsibility for the well-being of the whole biosphere, not really knowing what we are doing, we will as certainly produce havoc under the restorationist regime as we did under the aegis of domination. We could profit from the salutary reminder that the projects of domination spawned by the old model were all boosted with extravagant promises of beneficial results, were motivated by what were considered the noblest intentions, and were backed by the best science of the day—just as the new programs of control are touted now. For these reasons the restorationist rejection of the old domination model seems not nearly complete enough to restore a healthy relation between humans and nature. From this perspective, the restorationist shift to a new end in view does little more than dress up the old domination program in a currently fashionable green.

None of the foregoing implies that actual efforts of restoration should not be undertaken. This essay is not focused on ground-level projects of restoration but on philosophical principles in terms of which restoration is conceived and justified. If the philosophical principles of Jordan and Turner are defective, that does not entail that particular restoration projects are unsound. Individual projects—just because they are ground-level and do not reach out to encompass the entire biosphere—may have a potential for good that the philosophies used to justify them do not. Indeed, it seems to me that in our own day restoration is an inescapable obligation. Not, however, for the grandiose reason that we have ultimate responsibility for the health and well-being of the biosphere, but on the more homely grounds that when we make a mess we should do what we can to clean it up. This is more pedestrian and less exciting, but much more befitting members of a community that share the same neighborhood with others.

Restoration and Maker's Knowledge

One of the signal advantages claimed for restoration by its enthusiasts is that it gives us the highest kind of understanding. The idea, as Jordan explains, is that we understand something to the degree that we can assemble and control it (Jordan et al. 1987, 12; Jordan, " 'Sunflower Forest,' " this volume). As the practice of assembling (or reassembling) ecosystems, restoration provides us at once with the fullest knowledge and the ultimate test of our understanding of them. This involves more than just knowing the actions that will in some particular circumstances result in bringing about restoration, for restoration accomplished in this fashion may be no more than a case of someone triggering natural processes that then, apart from anything he or she knows or controls, get the job done. Restoration in its most thoroughgoing sense is a matter of full control, and only the most complete knowledge gives that to us.

In certain ancient disputes about knowledge, a distinction was drawn between beholder's knowledge, user's knowledge, and maker's knowledge; and a question was asked, triggering a debate that continued into the modern era, about who understands a thing the best: its maker, its beholder, or its user (Perez-Ramos 1988, chap. 5). Restorationists take their stand unequivocally with maker's knowledge.

This stand has become almost an article of faith in the modern scientific West. It can no longer be taken for granted, however, for the question of the supremacy of maker's knowledge is an issue we are forced to reopen as part of the profound rethinking of basic assumptions prompted by the environmental crisis. We may start a reexamination by noting that in many cases it is highly questionable that maker's knowledge provides us the best understanding. Take pain and torture, for example (Perez-Ramos 1988, 51). Who understands better what these are: the person who causes them or the person who suffers them? Or a child: does the physician who produces the baby through in vitro fertilization understand this child better than the parent who loves it, cares for it, and nurtures it to maturity? Or even consider a technical instrument—the computer, let's say—the type of thing for which the claim for maker's knowledge seems most compelling. Does the one who designs and manufactures the computer really understand it better than the historian or sociologist who sees the wide variety of its uses and effects in the lives, the activities, the thinking, and the satisfactions of individuals, and in all the manifold and complex transformations it brings to the world of work, to business, to social and economic structures, politics, international relations, military policy, education, science, demographic patterns, human self-understanding, philosophical worldview, cultural ideals, and so on? In each of these instances it would be difficult to make out the case for maker's knowledge. This is not to suggest that maker's knowledge never provides the best understanding, but only to show that whether maker's knowledge really gives us the best understanding of a thing is related to two factors: what kind of thing it is, and how it fits into a broader context.

Another point to note is that the kind of knowledge we can have of anything is influenced by the relation we take to it and the purpose we have for it. This is evident in the fact that the three forms of knowledge in question so far are named after relations. We can easily imagine other relations that yield their own forms of knowledge, such as lover's knowledge, disciple's knowledge, parent's knowledge, enemy's knowledge, worshiper's knowledge, partner's knowledge, and so on. There is a different form of knowledge for every different relation a would-be knower may have to what he or she wants to know.

Combining the results of the two preceding paragraphs, we are led to the conclusion that the kind of knowledge that constitutes the deepest understanding of something depends on what the thing is, how it fits into a larger picture, and what our proper relation to it is. When we apply this conclusion to nature—to the biosphere and its systems and subsystems—we are brought back to the topics of the previous section, the nature of nature and the character of our relation to it. The same issues are at stake both here and there, only now we are looking at their epistemological side instead of their metaphysical side. Whether maker's knowledge can provide us the best understanding of living ecosystems depends on whether the biosphere is best represented as a community or by some other model.

These themes have already been presented in their metaphysical guise, and now it may be valuable to point out some epistemological equivalents. First, because maker's knowledge, and its close associate controller's knowledge, are the epistemology of the old domination program, it's hard to see how the restorationists' conception of knowledge takes them away from this in any significant way. To be sure, restorationists stress the importance of ecological relationships, but that in itself does not represent abandonment of domination as the goal; it may signal only an adjustment of the old domination program to the new order of nature disclosed by the ecological sciences. Jordan argues, for example, that forms of restoration that proceed by making a few key adjustments and then allowing nature to take its course are not the most valuable, as "the essential idea is control—the ability not only to restore quickly, but to restore at will, controlling speed, altering its course, *steering* it, even preventing it entirely . . . " (Jordan et al. 1987, 17; emphasis in the original). He also writes that full understanding of a system can be gained only when we see what happens when it is disturbed, and, he adds, "the perturbations must be extreme" (1987, 12). What is this if not a continuation of the Baconian domination science that studies "nature under constraint and vexed," on the premise that "the nature of things betrays itself more readily under the vexations of art than in its natural freedom" (Bacon 1960, 25)? Indeed, domination is so much the overriding objective of the research Jordan describes that the actual restoration of nature is not a final end after all, but a means to gaining control.

As we saw earlier, this does not fit in with what seems to be the best model of nature, because domination and control are radically contrary to the spirit of community and violate the independence and autonomy of its members. What we need to note here is that independence and autonomy have their epistemological counterpart, in mystery. Any entity marked by a degree of autonomy has within it a level of being that is inscrutable. It has a dimension of itself, an inner reality, a unique element of what it is to be the thing it is, that is inaccessible to anything else, perhaps even (to some extent) to itself. It has an identity, a value, a purpose, a creativity that escapes reduction to the categories of objective knowledge and to the uses of human beings. Such a dimension, though not beyond destruction, is beyond mastery (including intellectual mastery) and control by others. Perhaps the fact that as our knowledge of nature has grown ignorance has grown apace (Ravetz 1986) is eloquent testimony to the element of mystery at the heart of things. If humans indeed live in a great community of being, our lives are governed by mystery more deeply than they are governed by knowledge, and we need an ethic of the mysterious—appropriate means for approaching and responding to mystery and autonomy—more than continual increase of maker's knowledge. (An ethic of the mysterious is of a piece with the ethic of trust discussed earlier.)

A good beginning in the development of such an ethic is the recognition and preservation of places where the autonomy and mystery of things is respected, and where creatures are allowed to be themselves, without the disturbing intrusions of those who would dominate them. People coming into such places will set aside the quest for knowledge and control that has become normal today, and they will find opportunity instead for quiet and undisturbed encounter with the mysteries. In these places we may open ourselves up to the wonder of the great community of being, we may admire the marvelous forces of Evolution or God (or both) that have brought forth the intricate structures of our world and its biosphere, and experience the paradox of being at once humbled and uplifted. In the process we may gradually grow in our ability to see and hear and understand, not as controlling knowledge understands, but in ways that beings who care for each other and depend on each other—parents, children, friends, lovers—understand. This experience has the capacity to fill, inspire, change, and invigorate people—

and empower them for a return to daily activity, bearing a sense of the awesomeness of Life and Being that gives meaning and definition to everyday pursuits while at the same time placing the needed constraints upon them.

Legally designated wildernesses are not the only possible places of this type. They certainly have their importance in serving the values indicated, and they may be particularly crucial as refuges for species that do not get along well with humans in the same habitat (Callicott 1991, 236). But they tend to be far away from the centers of human population. We also need many smaller areas scattered here and there—especially in our cities and suburbs, where undisturbed nature tends to be squeezed out of the environment, and where they would be more accessible to the average city dweller. In addition, we need places in our own souls, and in the whole range of the relations we maintain, that have not been submitted to the procedures and the uses of maker's knowledge, and which can serve as points of contact with a level of being and a richness of life that is not of our making or our control (Berry 1987). If wholeness in the nature of things and in ourselves is to be recovered, we need these points of contact everywhere.

We are now in a position to see why the tone of the criticism that restorationists direct against preservation is so harsh and visceral. Because initially it strikes one that there is no incompatibility between restoration and preservation, and that the two could well join hands in a powerful environmental alliance, it is tempting to seek explanations in ad hominem terms: as a matter of polemical style, competition for public support, or even congenital dyspepsia. The issue goes deeper than that, however. By holding that humans are the lords of creation, restorationist metaphysics tolerates no enclaves anywhere kept free of human domination and control; by maintaining that maker's knowledge gives us the best understanding of things, restorationist epistemology tolerates no mystery. But mystery and the preservation of places kept free of human domination are at the heart of the preservationist program and philosophy. Preservation thus presents a direct challenge to the most fundamental assumptions and aims of the restorationist program, and vice versa. The disagreements between the two groups go all the way down to the foundations, to the level of basic philosophies concerning who we are, what nature is like, and what human life is about. This being the case, short of philosophical conversion on one side or the other, the disagreements are irreconcilable.

REFERENCES

Bacon, Francis. 1960. *The New Organon*. Library of Liberal Arts. Indianapolis: Bobbs–Merrill Educational Publishing.

Berman, Morris. 1981. *The Reenchantment of the World*. Ithaca and London: Cornell University Press.

Berry, Wendell. 1987. "Preserving Wildness." In Wendell Berry, *Home Economics*. San Francisco: North Point Press. 137-51.

Callicott, J. Baird. 1991. "The Wilderness Idea Revisited: The Sustainable Development Alternative." *The Environmental Professional* 13 (3): 235-47.

Devall, Bill, and George Sessions. 1985. *Deep Ecology*. Salt Lake City: Peregrine Smith Books.

Jordan, William R., III, Michael E. Gilpin, and John D. Aber. 1987. "Restoration Ecology: Ecological Restoration as a Technique for Basic Research." In Jordan, Gilpin, and Aber, eds., *Restoration Ecology: A Synthetic Approach to Ecological Research*. Cambridge: Cambridge University Press. 3-22.

Katz, Eric. 1985. "Organism, Community, and the 'Substitution Problem.'" *Environmental Ethics* 7 (3): 241-56.

Lovelock, James. 1988. *The Ages of Gaia: A Biography of Our Living Earth*. New York: W. W. Norton & Co.

Merchant, Carolyn. 1980. *The Death of Nature: Women, Ecology, and the Scientific Revolution*. San Francisco: Harper & Row.

Perez-Ramos, Antonio. 1988. *Francis Bacon's Idea of Science and the Maker's Knowledge Tradition*. Oxford: Clarendon Press.

Ravetz, J. R. 1986. "Usable Knowledge, Usable Ignorance: Incomplete Science with Policy Implications." In William C. Clark and R. E. Munn, eds., *Sustainable Development of the Biosphere*. Cambridge: Cambridge University Press. 415-32.

Roszak, Theodore. 1973. *Where the Wasteland Ends*. Garden City, N.Y.: Anchor Books.

Turner, Frederick. 1985. "Cultivating the American Garden: Toward a Secular View of Nature." *Harper's Magazine* 271 (August): 45-52.

———. 1988. "A Field Guide to the Synthetic Landscape: Toward a New Environmental Ethic." *Harper's Magazine* 276 (April): 49-55.

Humans Assert Sovereignty over Nature

Carl Pletsch

One place to locate the ideas of ecological restoration and invention is in the tradition of political theory. Strict adherence to a policy of preserving nature from human dominion represents an assertion of nature's independence. Landscape restoration and invention, however, seem to entail the extension of human sovereignty over nature.

The theory of popular sovereignty is only three hundred years old, but it has become one of the great *idées reçues* of our time. In the seventeenth century God was the ultimate source of all power, and kings ruled by "divine right." Popular sovereignty was put forward in seventeenth-century England as an alternative to divine-right monarchy. As introduced by John Locke in his *Two Treatises of Government* (1689/90), it served to legitimate the Glorious Revolution of 1688 and constitutional monarchy generally. It was later used to justify democratic revolutions in America, France, and elsewhere in the eighteenth and nineteenth centuries. Now one simply assumes that governments should serve the interests of the governed (and not vice versa), and that human communities and not the gods are responsible for their regimes. Most of the world's citizens and the constitutions of virtually every modern state subscribe to this dogma.

Popular sovereignty has not only conquered the minds of the world, it has become an ever more inclusive theory. Originally a political idea, since the French Revolution the idea of popular sovereignty has been extended to include the people's right to define a particular order of society, one incorporating legal equality and protecting certain "natural" rights. After initial efforts by the sansculottes during the French Revolution, revolutionary workers and socialist theorists throughout Europe in the nineteenth century succeeded in extending the idea of popular sovereignty to include the expectation of economic well-being as well as political and social welfare. And since World War I, virtually all governments have come to acknowledge a responsibility to maintain certain political, social, and economic standards for the people they represent. "The people" are now the presumed sovereigns in all of these areas. Popular sovereignty has thus come to cover almost every aspect of human culture. But so far the theory of popular sovereignty—which might also be called "human sovereignty"—has not been applied to our relationship with nature.

The human species has long exercised de facto sovereignty over nature, reshaping nature to its own purposes. This has been especially true since the scientific revolution of the seventeenth century so enhanced the human understanding of nature and thus also its capacity to manipulate nature. Locke's contemporary Isaac Newton might even be said to have ushered in the era of practical human sovereignty over nature with his synthesis of the laws of motion in *Principia Mathematica* (1686/87). Yet in spite of all the *practical* authority that modern science and technology has given humans over nature, a millennial convention *in theory* has exempted nature from the gradually expanding domain of human sovereignty. Therefore, human beings have lacked a theoretical, cultural, or normative basis upon which to develop a sense of responsibility to "husband" or care for nature and the earth. They have felt free to use nature without much regard for the consequences.

The distinction between "nature" on the one hand and "culture" on the other is one of the most ancient and fundamental oppositions in Western civilization. Felix von Heinimann's book *Nomos und Physis* (1945) explores the origins and significance of this resilient pair of categories. Clarence L. Glacken's monumental compilation, *Traces on the Rhodian Shore* (1967), shows how the original Greek dichotomy of nature and culture survived both Roman and Christian influences. In the modern era, even Darwinism, which so obviously relocates the human *in* nature, rather than outside and opposed to it, has been unable to reform popular thinking on this cleft of thought. Thus while humankind has been constantly enhancing its practical ability to dominate nature, the nature/culture distinction has perpetuated a belief in the intrinsic independence of nature.

Nevertheless, in the late twentieth century we commence planning the future of nature. Studies of environmental degradation and resource depletion, the growing preoccupation with "sustainability," and practical projects in restoring prairies, savannas, and tropical forests are all signs of this. Fantasies of transplanting our nature to other planets and improving upon nature here on earth are even more dramatic indications. The audacious titles to Frederick Turner's essays in *Harper's Magazine* are revealing: "cultivating" nature as our "garden" and creating a "synthetic landscape." The critical point is that, as we plan for the future of nature, we begin to assume responsibility for nature. As we abandon the idea that nature is independent, we begin to assert a theoretical human sovereignty over nature.

This is surely a dramatic moment, philosophically as well as practically. "Humans Assert Sovereignty over Nature" is a headline that will elicit two reactions, often in the same heart. One may recoil in horror at the thought that nothing could be sacred anymore if we no longer respect the autonomy and "otherness" of nature, but one may also sigh with relief that humans will finally take responsibility for what they have been doing to nature all along, and will begin to rectify their behavior in nature.

The first reaction has deep psychological and philosophical roots, but it is an unexamined habit of thinking, another *idée reçue*. As the one category believed by modern, secular people to be independent of human dominion, nature has been the residual sacred, the ground and guarantee of all that is beautiful, true, and good. Since the eighteenth century nature has been the ultimate source of human laws and norms. Whatever has been categorized as "natural" has been good. This view was reinforced by the fact that nature's history was so long by comparison to human history as to seem eternal. In this habit of thinking, nature is presented to the mind as reassuringly self-regulating, a deep and repeating constant by comparison with the changeable world of human culture. The writings of James Lovelock, as original as they may otherwise be, are just the most recent variation on this ancient prejudice. Normatively, therefore, nature was granted a kind of sovereignty over humankind. Because we have thought in this particular way for centuries, and because there is no other source of norms on the horizon,[1] it is terrifying to acknowledge that we may have to learn to think differently. Yet we may have to give up nature as the cornerstone of all meaning. That seems to be entailed in asserting human sovereignty over nature.

Learning to manage nature may be an unavoidable responsibility of humankind. How else can we reconcile our practical ability to alter and destroy nature with the biological imperative to perpetuate the human species? But if humans assume this responsibility and place themselves above nature, whence will they draw the norms and values to guide their husbandry?

This question has never arisen because people have heretofore not understood that humanity threatened nature as a whole. Sensitive observers throughout the centuries, of course, have recognized the deleterious influence of humankind upon portions of nature. In the late eighteenth century, for example, an English country parson named Gilbert White made a record of what he feared might be lost to encroaching industrial civilization in his village of Selborne (White 1789; Worster 1977, 2-25). Since then, and until very recently, however, it has seemed sufficient to conserve and protect nature. Nature was resilient enough that preventing certain human alterations of it, or at most keeping humans out of parts of it, seemed enough to guarantee its continued existence. Nature itself did not seem threatened. Hence the long delay in acknowledging the human role in altering nature: the sense of innocence derived not from a lack of knowledge about the human impact upon nature (that was well known), but from the unquestioned premise that nature as a whole was independent of human culture.

Now a few voices proclaim "the end of nature" and others foretell such a destiny. Now, in an emergency, we admit our role in corrupting nature and begin to plan drastic remedies. Thought-projects for transporting terrestrial nature to Mars belie the sense of emergency that derives from the belated recognition that nature as a whole is threatened. Injunctions to restore nature implicitly acknowledge that nature is *not* independent, just as Locke's argument that the people had restored the English monarchy by appointing William king underscored the sovereignty of the people. In this situation, we are already in one sense "beyond preservation."

The second reaction to the statement "Humans Assert Sovereignty over Nature" is therefore both the wiser and the more immediately practical one. We must learn to play a more constructive role in nature. Having already altered nature immeasurably, having brought about countless extinctions and permanently altered significant portions of the earthly biosphere, humankind has obviously been exercising an anarchic, selfish, and destructive sovereignty over nature for a long time. To take responsibility for nature is only appropriate, and now is none too soon. Even in order to guarantee life to future generations of the human species, people must learn to assert some form of hopeful, reasonable, and collective sovereignty over nature. There is no alternative.

This is a paradoxical situation fraught with danger. Human beings have been driven by their own destructive capacities to acknowledge their role in nature and begin to accept some responsibility for it, but our species is yet ill equipped to deal with this responsibility either theoretically or morally. Realizing that we are beyond that point in history where preservation seemed adequate, we are finding it difficult to define and settle upon a new ecological paradigm for relating human activity to nature. This is the need that Frederick Turner, William Jordan, and others are attempting to answer in the most creative ways. But as difficult as creating a new ecological paradigm may seem, that may be the easier task. More difficult will be discovering or inventing a new source of norms for managing nature as well as our relationships among ourselves. Since the death of God in the eighteenth century, nature has been our source of norms; now as we assume responsibility for nature and assert our sovereignty over nature, we are about to deprive ourselves of our only remaining external source of guidance. This will be a revolution no less daunting than the Copernican and Darwinian revolutions. Morally speaking, asserting sovereignty over nature will leave the human species utterly alone in the universe.

NOTES

1. Some will propose as another source of norms one or another traditional religion, many of which prescribe respect and reverence for nature. But that is to ignore that during the past several hundred years the world has been ruled in ever more secular terms.

REFERENCES

Glacken, Clarence L. 1967. *Traces on the Rhodian Shore: Nature and Culture in Western Thought from Ancient Times to the End of the Eighteenth Century.* Berkeley: University of California Press.
Heinimann, Felix von. 1945. *Nomos und Physis. Herkunft und Bedeutung einer Antithese im griechischen Denken des 5. Jahrhunderts.* Basel: Friedrich Reinhardt.
Locke, John. 1689/90. *Two Treatises of Government.* London: Awnsham Churchill.
Newton, Isaac. 1686/87. *Philosophiae naturalis principia mathematica.* London.
Turner, Frederick. 1985. "Cultivating the American Garden: Toward a Secular View of Nature." *Harper's Magazine* 271 (August): 45-52.
_____. 1988. "A Field Guide to the Synthetic Landscape: Toward a New Environmental Ethic." *Harper's Magazine* 276 (April): 49-55.
White, Gilbert. 1789. *The Natural History of Selborne.*
Worster, Donald. 1977. *Nature's Economy: A History of Ecological Ideas.* New York: Cambridge University Press.

Landscape Restoration: More than Ritual and Gardening

Gene E. Willeke

Commenting on the essays by William Jordan and Frederick Turner that open this volume presents significant difficulties. Jordan's assumptions about what environmentalists are and what they think are more like a straw man than a real portrayal. He seems to equate environmentalists and preservationists, and considers preservation impossible. Although some environmentalists are strict preservationists, their numbers are few. Moreover, even the preservationists have no illusions about the pristine character of land that has been set aside as wilderness. Human influence is widely acknowledged, and when a tract of land is set aside as wilderness, it normally must meet criteria that are intended to ensure that the land has been impacted as little as possible by gross human activities such as logging, grazing, and mining. Of those lands that have been set aside in this manner, the impact of human activity has probably met these criteria, because such areas tend to be in places where there are major difficulties in exploiting the resources. One of the reasons for this kind of preservation is to provide models and samples (admittedly crude and incomplete) of the supposedly untouched land, as well as models for restoring degraded land.

Turner defines art as the creation of beauty, a narrow conception of art that is at odds with the goals of many artists who conceive of art as a means of communication and reflection. The art emerging from the creative process thus may be ugly as well as beautiful.

The term gardening is highly favored by both Jordan and Turner. This also is troubling inasmuch as gardening implies manipulation to achieve narrow ends, continuous tending, and the conscious removal of "weeds." Restoration should mean much more than gardening; it has such a vital role to play in the world that a more satisfying basis than only gardening and ritual is needed.

Restoration of a portion of a landscape has technical, social, and moral dimensions. For many people, interest in restoring landscapes derives ultimately from the destruction of these same landscapes. Soil erosion, replacement of a diverse ecosystem with a monoculture, clogging of a river by diseased, fallen trees, and extreme invasion of an ecosystem by a "weed" or an "exotic" species are examples of conditions that could lead to a desire to engage in a restoration activity. The term restoration strongly implies a return to some preexisting condition that is considered preferable to the present one.

A novel aspect of Jordan's discussion of prairie burning, and, indeed, of much of his approach to restoration, is the concept of ritual in restoration, which is elevated to the status of a religion, a position I find quite unappealing. The importance of ritual to Jordan seems to stem from a belief that only through ritual can humans achieve a satisfactory relationship with the nonhuman natural world. Again, this sounds like a straw-man argument, implying that there is only one way, or at most a limited number of ways, for humans to have healthy interaction with nature.

This discussion implies that if ritual is not present, restoration isn't occurring. One wonders, then, how a nation with an abundance of devastated land—Ethiopia, Haiti, Brazil, the United States—can seriously think of large-scale restoration if ritual must be a part of it. However satisfying the ritual may be, those who are interested in the ritual and are willing and able to participate are probably not sufficiently numerous for the task. One gets the impression from Jordan's essay that perhaps ritual can be symbolic and token in such situations, and that only a few need engage in it before turning the project over to others to do for pay. In a sense, the "restoration priests" are making public penance for the sins of the many, and are performing a cleansing ceremony. The priests might not even have to become involved in much of the technical work of doing the right thing, as long as what was done wasn't demonstrably the wrong thing and gave ample recognition to the need to change from the ways of the past to the ways of the future: restoration, rebuilding, and making whole.

The moral dimension of restoration is closely tied to the moral code and the beliefs of the leaders, who may be the priests or who may be the ones who provide the technical expertise. Jordan seems to make a distinction between the priests and those with the technical background to understand more of the complexities of ecosystems. "Moral" has something to do with rightness. Restorers want to do the right thing. Here is a source of trouble, but it cannot be avoided. How should something be restored? Whose values should prevail? Should sustainability be a highly ordered value? Or diversity? Or economic worth? Or beauty? Or orderliness? Or simplicity? Or naturalness?

For the sake of argument, and I believe these choices are reasonable for the United States in the 1990s, consider the impact of exalting to a place of primacy the older values of beauty, orderliness, and diversity, and the newer one of sustainability.

Beauty is much prized in our society, especially now that a large proportion of the population, and virtually all of the "decision-making" members, are able to satisfy fundamental human survival needs. It is well known that what is beautiful to one may not be beautiful to another. Beauty, however defined, may be gained at the expense of diversity and sustainability.

Orderliness in restoration implies human control. This does not mean that there isn't order in either the universe or in ecosystems, but it does imply that in seeking orderliness in restoration a different aesthetic is employed from what may exist in the preexisting natural state. This aesthetic may try to impose certain patterns or to enhance the roles of color or texture or symmetry in restoring a landscape. Orderliness may be achieved at the cost of constant maintenance.

Diversity is valued, and then exploited, in part because of a belief by some people that does not have much supporting empirical evidence: that biological diversity leads to stability and sustainability. In a natural system this may well be, but it is less obvious when the diversity is imposed, without a full understanding of the consequences, in a nonevolutionary manner. Diversity may be increased by the introduction of one or more species, but the consequences may be the obverse of stability and sustainability. Plant diseases have resulted from the introduction of an exotic species that may threaten species native to the restoration area.

Sustainability is, at this juncture, a somewhat nebulous term, praised by some and derided by others who equate it with continued exploitation. The term implies an ability to continue "production" at a roughly stable level for an indefinite period of time. Like the other valued traits, sustainability may be achieved at the cost of losing something else. To give an example, a piece of land named the Bachelor Reserve was bequeathed to Miami University, with the proviso that diversity of wildlife should be maximized. This innocently stated value, written in the will of a humanist and lover of the outdoors, has led to the planting of vegetation to provide food and shelter for wildlife. Some of these plants would not have been on the reserve in its natural state, or would not have been present in such numbers. Because humans are in control of this ecosystem, however, these things can be, and have been carried out. The only successful planting, in terms of the survival of the plantings, was a conifer plantation; other attempts at planting have not been successful (though some annual plantings of feed crops were done to enhance animal diversity). The results have tended to fall most readily into the categories of orderliness, diversity, and beauty. Simplicity and naturalness have been less valued.

A colleague of mine is a respected practitioner of river restoration in the United States. He has been able to work with great artistry on streams clogged with log and debris jams, obstructed by sediment, and whose banks have been seriously eroded. In restoring rivers, he employs a strategy of obstruction removal, flow deflection with river training works made of natural materials found in or near the river, and revegetation of banks to minimize soil erosion. His skill is accompanied by a set of values, in which particular species of trees are more valuable than others and some planforms are preferred over others (oxbows, for example, are considered a waste). Certain trees should be cut because they "aren't worth much." The look of naturalness is more important than real naturalness.

Some restoration specialists believe they can improve on the natural state or condition, which might give rise to a statement such as "God did well, but no one should seriously believe that human beings can't improve now and then on creation." It is not hard to see how such an attitude could develop, as a restoration specialist might spend years attempting to understand an ecosystem and then more years perfecting methods to build such ecosystems. This also fits the job description of a restoration priest. However, ecosystem complexity is such that in building ecosystems, only a few elements can be consciously manipulated with ease, and those elements tend to have strong positive emotional appeal. (The grasses of the prairie, for example, are often treated rhetorically with the kind of language typically reserved for beautiful flowers and majestic trees.) One might be more willing to accept the validity of such an ecosystem restoration philosophy if nettles and poison ivy were reintroduced into floodplains, mosquitoes and deerflies were reintroduced on northern lake shores, and copperheads into southern forests. These organisms do not have widespread positive emotional appeal, but they are parts of the ecosystem, and a commitment to restoration should probably be a commitment to as complete a restoration as possible.

The person holding an extreme view in the complex of restoration positions, one that extolled naturalness above all else, faces different dilemmas. Someone with this position who is serious about restoration might be forced either to allow succession to take place over centuries by eliminating human influence as much as possible or to concede that naturalness cannot be attained, and therefore restoration is futile.

In imposing a set of personal or group values on a landscape restoration program, certain risks are encountered, and these risks are moderated by the extent of scientific knowledge about the behavior of ecosystems and the skill and artistry of those doing the restoration. If one makes the reasonable assumption that no one doing restoration would knowingly do something truly destructive, then the risks are narrowed, and should, over time, become increasingly narrower. The "should" is an important qualifier, however. Had the potential impact of introducing kudzu into the southeastern United States, or multiflora rose in the midwestern states, been anticipated, these decisions likely would have been different. Both plants were introduced with conservation goals in mind, but those introducing these species were not fully aware of the potential impacts. In the absence of such knowledge, and in a flush of excitement at being able to bring an organism into an environment that had great promise, these species were promoted vigorously. It might be argued that without the promotional fervor, the impacts would have been less severe, but less promotion of these species may have only delayed the inevitable spread of plants that came to assume the properties of pests.

These cases may be extreme. The introduction of other species or the failure to reintroduce native species, for whatever reason, need not and often does not result in ecological catastrophe, even though there may be adverse impacts. Ginkgo trees, nonnative to the United States, have not been disastrous, and other similar examples of plants and animals could also be given.

I find it surprising that neither Jordan nor Turner mentions succession as an appropriate restoration strategy. My river restoration colleague makes considerable use of succession. When he undertakes revegetation of the riverbanks, he starts with willows or similar species that will grow in an early successional stage and help re-create conditions that will allow other species to become established, which ultimately will be more important in holding the soil and providing shade for the channel. Succession would seem to be a less labor-intensive strategy than the kind of gardening discussed by Jordan.

Having engaged in a fairly harsh critical appraisal of the Jordan and Turner essays, I would like to conclude by saying that one need not subscribe to the notions of ritual and gardening to profit from the experience of restorationists like Jordan. The approaches and techniques developed to restore particular ecosystems can be used effectively without the trappings of ritual and without advocating that restoration is the same as gardening.

It is also possible to accept Turner's premise that artistic talent, or his "aesthetic sense," represents "a capacity to organize and recognize meaning in very large quantities of ill-defined information, to detect and create complex relationships and feedback systems, to take into account multiple contexts and frames of reference, and to perceive harmonies and regularities that add up to a deep unity." The complexity of ecosystems is such that artistic talent and insight are, for the foreseeable future, among the promising avenues for restoration, particularly when the insights and data of science can be married to this aesthetic sense. I have described my river restoration colleague as an artist from the first time I observed his work and listened to him explain his methods. He is willing to accept scientific findings that are not inconsistent with his observational experience of river ecosystems. If there is an inconsistency, he tends to conclude that the science must be bad or undeveloped, because it does not jibe with his conception and experience of reality. He happens to fare better in his river restoration work than do many whose scientific competence is greater but still inadequate to the task without the supplemental insights provided by what Turner calls "aesthetic sense."

In summary, the restoration process is vital and must be practiced widely throughout much of the world, even though it is hardly necessary to refer to it as a "new environmental paradigm." Restoration as ritual will not likely become an important element of the restoration process. Healthy relationships between human beings and the natural landscape can be developed in other ways more compatible with their worldviews and patterns of life. If the world is to retain or, often, regain its productive capacity, restoration of vast tracts of land will be necessary. Many of those involved in this work will find it deeply satisfying, and certainly not everyone who works on restoration must have the same aesthetic or spiritual experience. Both Jordan and Turner are to be commended for developing these ideas about restoration and presenting them for the consideration of a thoughtful community who, however critical of portions of the argument, has deep respect for the authors and the subject matter and great hope for the outcome.

Changing Worldviews and Landscape Restoration

Dora G. Lodwick

Landscape restoration is an idea "whose time has come" only if supported by a broad set of beliefs and values about how people relate to nature. Americans have struggled to define this relationship since the end of the nineteenth century, as it became apparent that the "vast frontier" of nature would eventually be filled unless something were done to protect it. Perceptions about human relationships to nature have varied depending on the social class and racial-group characteristics of individuals and depending on the nature of the dominant worldview of particular historical periods. Therefore, placing Frederick Turner and William Jordan's call for landscape restoration within the historical context of the U.S. environmental movement and its changing worldviews will be helpful.

Landscape restoration is a modern version of the early preservationist and conservationist strains of the American environmental movement. These positions have been held by a specific group of people in our society . . . the well-to-do. The growing perception that the environment is creating serious threats to human health and safety (Dunlap 1991), which has become widespread in the United States in the 1980s and 1990s, is especially focused among people of color and those with working-class backgrounds (Bullard 1990). Both social-class-based perspectives about the relationship of people to the natural environment are supported by the nation's changing worldview that has become dominant since the 1980s and which stresses the interdependence of humans with nature.

Historically, the upper and upper-middle classes in industrialized nations have had greater effects on nature than the lower classes. These effects have been created through the elites' decision-making positions in government, industry, and environmental movement organizations, and through their consumption choices (Bullard and Wright 1992; Morrison and Dunlap 1986). The tendency of Turner and Jordan to focus on human influence on nature reflects this social class reality. I propose that they need to expand their own perspective to include the survival issues of broader social classes and racial groups that are concerned more about the effects of nature on their ability to survive and less on how they affect nature. Such a broad-based landscape-restoration program has implications for how people learn to trust each other across class and racial lines (see Kane, "Restoration or Preservation? Reflections on a Clash of Environmental Principles," this volume).

To broaden the landscape-restoration proposal, we need to address several concepts and assumptions about how human beings relate to their natural environment. First, the natural environment provides two basic functions for people: their habitat and their sustenance (Schnaiberg 1980; Logan and Molotch 1987). As social groups focus more intensely on either habitat or sustenance, they develop different perspectives about ecological and social realities.

Habitat social movement participants have historically considered the environment primarily as habitat for nonhuman flora and fauna. They have also supported the outdoor recreational use of nature. This focus on amenities has been held predominantly by members of the upper and upper-middle social classes (Schnaiberg 1980).

Sustenance social movement participants historically have been concerned with how industry uses natural resources. These social movement members have concentrated on conserving the natural resources of the nation, calling for energy preservation, recycling of resources, and the adoption of appropriate technologies, including soft energy technologies.

Lately both habitat and sustenance perspectives have merged with issues of equity and social justice, particularly as lower-class and minority racial groups have expressed dismay at the toxicity of their habitats. A disproportionate share of humanly created toxic environments has been placed in lower-class or racial minority communities (Bullard 1990; Bullard and Wright 1992). Forty percent of the nation's total hazardous waste disposal capacity, in 1987, was in African-American or Latino communities (Commission for Racial Justice 1987). These groups no longer accept sustenance (e.g., jobs) over safe habitat.

Historical Roots

One of the earliest American environmental social movements was the progressive conservation movement, which was concerned with the destruction of forests and recreational areas. It shaped attitudes about economic growth, pollution, and resources until about 1970 (Humphrey and Buttel 1982). The majority of the movement's members were from the upper and upper-middle classes.

The habitat and sustenance themes were apparent in two strands of the movement. Preservationists, like the Sierra Club and the Audubon Society, wanted to keep the habitat undeveloped so the public could learn from the areas and use them for recreation. Conservationists, such as large ranchers and miners, wanted to reduce the rate of natural resource use to "wisely husband" it for future human use, thus reflecting the sustenance emphasis (see Turner, "The Invented Landscape," this volume).

The progressive conservation movement pressured government to assume greater control over public land. Several legislative efforts were passed, setting aside national park areas in various states. The forest service and park service were created and promoted land reclamation and irrigation. This helped large corporations that were able to adopt technological innovations appealing to the "wise use" of natural resources (Hays 1959).

The modern environmental social movement integrated the habitat and sustenance themes of the progressive conservation movement in the 1960s, and continued its relationship with the government. Government agencies, especially the Environmental Protection Agency, the Department of Energy, and the Department of Interior, hired many impact scientists in response to some of the movement's developments (Schnaiberg 1980).

Corporate, national environmental, and governmental organizations have been the primary actors in the modern environmental movement. Meanwhile, public political actions have been mobilized predominantly through national environmental organizations and smaller, local interest groups (Schnaiberg 1980). New actors have emerged with the rise of the toxic-waste interest of the current environmental movement.

Public Opinion

Social problems appear and disappear on the American scene. Downs (1972) has identified a five-stage "issue-attention cycle," according to which environmental problems would go through the cycle, then move into a natural decline and eventually disappear from the social horizon. In fact, worries about environmental quality, from the mid–1960s to 1990, have failed to follow this cycle (Dunlap 1991; Dunlap and Scarce 1991).

Modern public concern with the environment developed very rapidly in the late 1960s. Rachel Carson (1962) and Paul Ehrlich (1968) were writing about how the growth in human society was causing an imbalance in the natural ecosystems. Prior to 1965, environmental quality was interesting mainly to air and water pollution officials and conservationists (Dunlap 1991), but from 1965 to 1970, air and water pollution went from being perceived as a significant issue by a small minority of people to being the second most important concern of Americans (Humphrey and Buttel 1982, 123). The National Environmental Policy Act (NEPA), passed in 1969, was the culmination of several environmental legislations passed by Congress during this period.

Environmental interest peaked around the first Earth Day, in 1970. The majority of the public declared proenvironmental opinions, and only a few people expressed antienvironmental attitudes. The general population, however, did not see the environment as one of the country's most important problems (Dunlap 1991).

The attention directed to the environment declined rapidly soon after Earth Day in 1970, then more slowly throughout the rest of the decade. At the end of the decade environmental issues still had greater support than in the mid-1960s (Dunlap 1991; Dunlap and Scarce 1991). For example, people were asked to identify the three national problems, from a list of ten, that most needed more governmental attention. "Reducing pollution of air and water" was selected by 17 percent of the population in 1965; by 53 percent in 1970; and by 24 percent in 1980 (Dunlap 1991, 297).

The 1980s experienced considerable increase in demand for environmental protection. Some of this appeared as a backlash against Reagan's environmental policies and the continual emergence of new environmental problems (Dunlap 1991; Dunlap and Scarce 1991). For example, in 1982, 35 percent of the U.S. population said there was too little governmental regulation and involvement in environmental protection and only 11 percent said there was too much. In 1986, 59 percent said "too little" and 7 percent said "too much"; in 1990, "too little" was selected by 62 percent (Dunlap 1991, 301). The public consistently has suggested that the government should be responsible for creating a safe environment.

The advocacy continued into the Bush administration with further emphasis on global warming, ozone depletion, and specific problems

such as the *Exxon Valdez* oil spill. The "increased awareness of the growing seriousness of ecological problems . . . has probably become the critical force in driving public opinion," concluded Dunlap (1991, 302). He
continues:

> There is a widespread perception that the quality of the environment—
> from the local to the global level—is deteriorating. Furthermore, this
> deterioration is seen as posing a direct threat to the health and well-
> being of humans. . . . This growing sense that environmental conditions
> are becoming truly threatening to our future provides a depth to
> environmental concern that was largely absent in 1970. (308)

Public support for environmental protection is now extremely high.
Public worries about environmental deterioration and the demand for action reflects the current dominant worldview.

Changing Worldviews

Societies develop worldviews, which change over time. The dominant
worldview is the belief and value system prevalent among the people of
a society at a particular time, providing some sense of integration for the
social fabric. Social groups will more easily find societal acceptance for
their claims if those claims are consistent with the dominant worldview.

The Industrial Worldview was dominant in the United States from the
period when the progressive conservation movement was strong (1890–
1920s) until very recently. Analyzing this worldview is important, because it provided the "lens" through which the early environmentalists
saw the world as they agitated for the conservation or preservation of the
land—both for habitat and for sustenance reasons.

For example, the members of the progressive conservation movement
pressured the government to control use of public land. The "uses" included land reclamation, irrigation, and "wise use" of natural resources,
all of which were seen as "natural and familiar" (Gamson and Modigliani
1989) because they were consistent with the Industrial Worldview, which
stressed controlling nature through science and technology to create continual economic growth. Therefore this dominant worldview of the time
added weight or "socially amplified" the claims of the progressive conservation movement (Machlis and Rosa 1990). These claims had legitimacy, and therefore power. The issues of the movement appeared natural
and familiar, thus acquiring "cultural resonance" (Gamson and
Modigliani 1989).

The primary tenets of the Industrial Worldview were the following (Olsen et al. 1992):

- belief in controlling nature to improve human life
- belief in the efficacy of science and technology
- belief in economic growth
- belief in centralized, hierarchical government
- belief in large organizations
- belief in individualism and competitive interactions
- belief in material abundance.

Ideas from worldviews extend throughout society the ideologies developed by particular social groups. An ideology is "an argument derived from a worldview or social paradigm that a group of people purposefully use to justify their actions" (Olsen et al. 1992, 19). The dimensions of the Industrial Worldview can be seen in the ideology of the progressive conservation movement. The movement's claims were consistent with the mastery of nature through human planning and control of the parks and forests of our nation.

During the 1970s, social scientists found increasing evidence that the dominant worldview in the United States was changing. Key dimensions of the subsequent Postindustrial Worldview were (Olsen et al. 1992):

- belief that humans are part of nature's ecosystem
- belief that science and technology create solutions and problems
- belief that natural resource constraints limit economic growth
- belief that government should be decentralized and nonhierarchical
- belief that small and simple organizations are best
- belief that collectivism should be encouraged
- belief in future material scarcity.

The Postindustrial Worldview challenges us to consider a number of questions. What is the nature of this worldview? How widespread is it? What is driving the change to a new dominant worldview?

Olsen et al. (1992) examined the internal relationships among the dimensions of the Postindustrial Worldview. After empirically documenting the loose consistency of these characteristics, they focused only on the central dimension of this worldview as captured in the ecological social paradigm. A social paradigm, a more limited aspect of a worldview, is defined by Olsen et al. as "the perceptual and cognitive orientation that a 'communicative community' uses to interpret and explain particular aspects of social life that are important to it" (18). Relationships between nature and people are the important aspects of social life for the "communicative community" concerned with landscape restoration.

The authors tested for the presence of the ecological social paradigm by conducting a mail survey of 696 adults randomly selected from the residents of Washington State who had listed telephone numbers in 1982. The usable responses reflected a 65 percent response rate. Because the sample was drawn only from Washington, the results could not be generalized to the entire United States, although they were very similar to results obtained in ten other states during the same time period and to questions fielded in a 1988 urban survey in Michigan (Olsen et al. 1992). This consistency increases the reliability of the results, suggesting that the findings are quite robust across time and place.

Like worldviews, social paradigms are composed of interrelated beliefs and values. The beliefs of the ecological social paradigm, as reflected in the responses of the survey participants, included:

- People must learn to live in harmony with nature to survive: 86 percent agreed.
- The earth is like a spaceship, with limited room and resources: 78 percent agreed.
- Modern industrial countries are very seriously disturbing the balance of nature: 78 percent agreed.
- Despite our special abilities, humans are subject to the laws of nature like other species: 62 percent agreed.

The values of the ecological social paradigm were revealed in the survey responses as follows:

- People should adapt to the environment whenever possible: 82 percent agreed.
- Natural resources should be saved for the benefit of future generations: 75 percent agreed.
- Nature should be preserved for its own sake: 55 percent agreed.
- Environmental protection should be given priority over economic growth: 51 percent agreed and 23 percent were undecided.

All the belief statements were strongly supported with over 60 percent agreement. The beliefs stress the interdependence of people with nature, which should socially amplify Jordan's "mutually beneficial" interaction claim for a healthy relationship between humans and nature (see Jordan, " 'Sunflower Forest': Ecological Restoration as the Basis for a New Environmental Paradigm," this volume).

Although the interrelationship was obviously accepted, how it should be developed was less clear, as demonstrated by the lower percentages of agreements (51-82 percent) for the value statements. The preservation argument received only a 55 percent agreement. When environmental protection was placed against the sustenance function of nature (e.g., economic growth), the result was barely a majority agreement (51 percent) and a high percentage of undecided (23 percent). This seems to suggest that landscape restoration that also considers human need for sustenance will have greater cultural resonance than the conservation argument.

The ecological social paradigm will provide greater social amplification only as it is part of a dominant worldview, so therefore the authors explored how widespread the ecological social paradigm was. They combined the values and beliefs, creating a typology of those who supported the paradigm. The authors found that 56 percent of the Washington sample held the ecological social paradigm either moderately or completely; 27 percent accepted it weakly or partially; and 17 percent clearly rejected the paradigm.

The only demographic characteristics associated with supporting the ecological social paradigm were age and education. Younger people of both genders were more likely to hold the paradigm than older people. (Age was specified in the intervals 18-35, 36-50, 51-65, and 66 years or older.) Persons with graduate educations were also likely to accept the paradigm; however, no further differences in the level of support for the paradigm were revealed among people at all lower levels of education (Olsen et al. 1992, 74). These findings were consistent with other studies (Dunlap and Van Liere 1983).

Because demographic characteristics did not seem to influence whether someone held the ecological social paradigm, the authors explored other variables to determine what was driving the change in worldview. Two theoretical arguments were derived from the work of Thomas Kuhn (1970). Kuhn's argument that internal logical contradictions between beliefs and values will drive paradigm changes was not supported by the data of this particular study (Olsen et al. 1992, 115).

The second argument, that external discrepancies between people's paradigms and their social conditions would fuel a paradigm change, was consistent with the findings. The variables that measured existing social conditions explained 33 percent of the total variance in the ecological social paradigm of this sample (Olsen et al. 1992, 136).

The variables used to measure social conditions included: (1) environmental movement cohort membership, (2) environmental scarcity awareness, and (3) environmental risks concerns. The environmental movement cohort membership variable was created by combining age (e.g., under or over 35 years old) with whether or not the respondent had a college education. The full cohort members were those who were 35 years old or younger and had at least a college education. Nonmembers of the cohort were those over 35 with no college education. Environmental scarcity awareness was created from questions about the perceived seriousness of the U.S. energy situation and of population pressures on natural resources. The environmental risks concerns variable was constructed from perceptions about how people negatively affect nature, future suitability of nature for humans, and potential radiation risks (Olsen et al. 1992, 116-26). This third variable was most strongly related to holding the ecological social paradigm in a multiple regression equation composed of risk concerns (beta $=$ 0.54), resource scarcity awareness (beta $=$ 0.17), and environmental movement cohort membership (beta $=$ 0.12).

The change to the ecological social paradigm, the central dimension of the Postindustrial Worldview, is therefore driven most strongly by the sample's perception of the risks to the habitat associated with current technological and scientific developments. Sustenance issues related to perceptions of resource scarcity are a weaker influence on the paradigm. This conclusion is consistent with the public opinion polls of the 1980s, which showed a growing concern with environmental threats. Public attention has remained focused on the quality of the environment because this is part of a broader change in worldview. The Postindustrial Worldview, with the ecological social paradigm, is becoming dominant in the United States in the late twentieth century, primarily because of perceptions of ecological threat.

Influence of Class and Race

The public-opinion data and the worldview data reflect the ideas of the majority of the U.S. population, but there is another group that has historically been silent about the environment, identifying more strongly with civil rights. In the past, lower-class Americans and African Americans have not organized around environmental issues because they have been predominantly concerned with sustenance. Most often industry and government have held out the promise of jobs as the trade-off for locating industrial facilities, which often create environmental problems, in lower-class or racial minority communities.

Currently no representative sample exists of the opinions of people of color about the environment, but a number of case studies have been conducted. Bullard's (1990) landmark study of five African-American and Latino communities provides an important contribution.

Bullard conducted a survey of 523 randomly selected households in five communities where problematic facilities had been sited. He found that 66 percent of the households rated concern for the environment as being more important than jobs. Overall, 62 percent of the households saw no improvement in employment of local residents as a result of the local facility (93). Bullard states: "Job blackmail has lost some of its appeal, especially in those areas where the economic incentives (jobs, taxes, monetary contributions, etc.) flow outside of the black community" (106).

A new strand of the environmental movement has developed around toxic contamination and equity (Bullard 1990; Bullard and Wright 1992). African Americans have generally defined their interest in the environment from their concern with social justice. Their community's institutions (e.g., churches, NAACP, Urban League, and so on) historically have fought for social-justice issues. Only recently have these organizations begun to link environmental disparities (e.g., inequitable distribution of clean environments) to institutional racism. This "has led black social action groups to adopt environmental equity as a civil rights issue, an issue well worth 'taking to the street' " (Bullard 1990, 108). A history of civil-rights strategies and tactics may reinforce African-American leadership in the toxicity environmental movements of the twenty-first century. Social justice and equity will become values socially amplified by the nonhierarchical dimensions of the Postindustrial Worldview. The threatening human habitat will override concerns about human sustenance.

Conclusion

William Jordan and Frederick Turner call for landscape restoration as a key to a mutually beneficial relationship between people and between humans and the rest of nature. They emphasize engaging all of human capabilities in the interaction with nature, then celebrating the relationship.

The time is right for this kind of relationship. The leadership that emerges during this period is critical to influencing the developments that will occur in the way humans relate to each other and to nature in the next thirty to fifty years. We are in the midst of a worldview shift. Olsen et al. (1992) have documented that a majority of their sample (56 percent) held the ecological social paradigm in 1982. This majority was even larger by 1988.

Dunlap (1991) suggests that concerns about the environment have not followed the short "issue-attention cycle" but instead have persisted. In fact, in 1990 interest in environmental issues was at the highest level it had ever been. The traditions and power of the civil-rights movement have also become engaged with the environment.

The American public wants something done about the perceived environmental deterioration. The public still shows strong support for governmental action to maintain environmental quality, for people are skeptical of individuals' and industries' protection of the environment (Dunlap and Scarce 1991, 655). Furthermore, 75 percent of Harris poll respondents in 1989 characterized environmentalists, from a list of six groups, as having done more good than harm (Dunlap 1991, 309). Only the Chamber of Commerce (selected by 76 percent) received a higher rating.

It is time to put forth leadership in showing how to interact with the environment. The change to the Postindustrial Worldview introduces a strong foundation of beliefs and values that will socially amplify and strengthen the claims of those who advocate a mutually beneficial relationship with nature. This will provide a stable cultural base that will legitimate and support the proposed landscape restoration approach.

To be able to take advantage of this historical moment, Jordan and Turner will need to broaden their position to include the active interaction of humans and nature in creating a secure habitat that also creates the necessary sustenance for humans and other species. The factors driving the worldview change are not the problems identified by Jordan and Turner. They claim the following to be responsible for the problems that we face: (1) a modern environmentalism that limits the relationship between humans and an inaccessible, idealized wilderness; (2) as a result, the inability of humans to use all of their senses and capabilities in relating

to nature; (3) the speed of cultural change—if by this they mean technological and scientific changes—that leads humans to feeling alienated
from nature; and (4) environmentalism's lack of optimism about the ability of humans to have a positive relationship with nature (see Jordan,
" 'Sunflower Forest,' " this volume). The writings of Jordan and Turner
reflect the roots of their environmental thinking . . . caught in the historical tradition of the progressive conservation movement. The idealized wilderness has been a concern primarily of the upper and upper-middle classes.

The American people are changing their worldview because of the
perceived risks that nature is carrying (Olsen et al. 1992; Dunlap 1991).
Nature is *not* seen as the pristine, distant wilderness of the elites. It embodies threats to the health and safety of humans who are simply trying
to live securely in their own habitat. Because of its web of connections,
nature has spread what humans have produced throughout the system.
Contaminated water, for example, was one of the major concerns in the
late 1980s (Dunlap 1991). Although Turner and Jordan suggest that the
speed of cultural change promotes a distancing from nature, this change
actually creates a *closer* relationship to nature.

Human habitat is no longer safe for humans; it is threatening, because
of previous human interactions with nature. To say simply that environmentalism has been too negative and not adequately celebrative does not
acknowledge the real threat being experienced by those living near toxic
contamination areas, municipal landfills, hazardous waste disposal areas,
and so on. People of minority racial groups and lower classes, and indeed
many in the general population of the United States, will not enter into
rituals that do nothing to clean up their habitat. These members of the
society do not feel separate from nature, but rather caught—too en-
meshed *in* nature.

Celebration can come only when there is trust, when there is security
in the habitat and sustenance functions of mutually beneficial exchanges
with nature. The American public has trusted government to help create
a safe habitat, but this trust has been shaken as the public discovers that
its habitat is no longer secure because of what uncontrolled humans have
done (Edelstein 1988; Levine 1982; Bullard 1990; Short and Clarke 1992).
In the 1990s, environmentalists may be the ones most trusted to provide
leadership in resolving some of the people's concerns (Dunlap 1991).

This leadership needs to start with the reality of the social context of
our society. Jordan and Turner must broaden their program to encompass

"protecting humans from the environment" (Bullard 1990, 117). They have already addressed the reciprocal—"protecting the environment from humans." Only when both are included can a mutually beneficial relationship exist.

Human capacity must be used to restore landscapes that have been so damaged that nature is not only hurt, but harmful to humans. These acts of community with nature and with each other will be mutually beneficial. They will free humans to trust again.

Ritual celebrations that begin before the environment is made safe for humans will be gaunt dances of the elites. A greater gulf will be created between people of different social classes and colors. Nature will be further alienated and alienating. This provides no mutual benefit to humans or to nature.

If landscape restoration embraces the challenge of expanding its vision to include a communal effort to address the environment's threats to human habitat and sustenance, as well as human threats to nature, then there will be cause to celebrate trusted, mutually beneficial relationships. This is the true challenge to Turner and human inventionists.

REFERENCES

Bullard, Robert D. 1990. *Dumping in Dixie: Race, Class and Environmental Quality*. Boulder, Colo.: Westview Press.

Bullard, Robert D., and Beverly H. Wright. 1992. "The Quest for Environmental Equity: Mobilizing the African-American Community for Social Change." In Riley E. Dunlap and Angela G. Mertig, eds., *American Environmentalism: The U.S. Environmental Movement, 1970-1990*. Philadelphia: Taylor and Francis. 39-49.

Carson, Rachel. 1962. *The Silent Spring*. Boston: Houghton Mifflin.

Commission for Racial Justice. 1987. *Toxic Wastes and Race: A National Report on the Racial and Socioeconomic Characteristics of Communities with Hazardous Wastes Sites*. New York: United Church of Christ.

Downs, Anthony. 1972. "Up and Down with Ecology—the 'Issue-Attention Cycle.'" *Public Interest* 28: 38-50.

Dunlap, Riley E. 1991. "Trends in Public Opinion toward Environmental Issues: 1965-1990." *Society and Natural Resources* 4: 285-312.

Dunlap, Riley E., and R. Scarce. 1991. "The Polls—a Report: Environmental Problems and Protection." *Public Opinion Quarterly* 55: 651-72.

Dunlap, Riley E., and Kent D. Van Liere. 1983. "Commitment to the Dominant Social Paradigm and Concern for Environmental Quality: An Empirical Examination." *Social Science Quarterly* 65: 1013-28.

Edelstein, Michael R. 1988. *Contaminated Communities: The Social and Psychological Impacts of Residential Toxic Exposure*. Boulder, Colo.: Westview Press.

Ehrlich, Paul. 1968. *The Population Bomb*. New York: Ballantine Books.

Gamson, William A., and Andre Modigliani. 1989. "Media Discourse and Public Opinion on Nuclear Power: A Constructionist Approach." *American Journal of Sociology* 95 (1): 1-37.

Hays, Samuel P. 1959. *Conservation and the Gospel of Efficiency*. Cambridge: Harvard University Press.

Humphrey, Craig R., and Frederick R. Buttel. 1982. *Environment, Energy and Society*. Belmont, Calif.: Wadsworth.

Kuhn, Thomas S. 1970. *The Structure of Scientific Revolutions*. 2d ed. Chicago: The University of Chicago Press.

Levine, Adeline. 1982. *Love Canal: Science, Politics, and People*. Lexington, Mass.: Lexington Books.

Logan, John, and Harvey Molotch. 1987. *Urban Fortunes: The Political Economy of Place*. Berkeley: University of California Press.

Machlis, Gary E., and Eugene A. Rosa. 1990. "Desired Risk: Broadening the Social Amplification of Risk Framework." *Risk Analysis* 10 (2): 161-68.

Morrison, Denton E., and Riley E. Dunlap. 1986. "Environmentalism and Elitism: A Conceptual and Empirical Analysis." *Environmental Management* 10: 581-89.

Olsen, Marvin, Dora G. Lodwick, and Riley E. Dunlap. 1992. *Viewing the World Ecologically*. Boulder, Colo.: Westview Press.

Schnaiberg, Allan. 1980. *The Environment: From Surplus to Scarcity*. New York: Oxford University Press.

Short, James F., Jr., and Lee Clarke. 1992. *Organizations, Uncertainties and Risk*. Boulder, Colo.: Westview Press.

PART III

PRACTICE

Restoration Ecology: Lessons Yet to Be Learned

Gary W. Barrett

Restoration ecology is a paradigm that emerged during the decade of the 1980s (Aber and Jordan 1985; Jordan et al. 1987; Jordan, " 'Sunflower Forest': Ecological Restoration as the Basis for a New Environmental Paradigm," this volume). Barrett (1989, 1992) noted that restoration ecology is one of several paradigms (e.g., landscape ecology, conservation biology, and agroecosystem ecology) in the area of applied ecology (Barrett 1984) that attempt to wed ecological theory with practical application. Restoration ecology is especially appealing to a broad array of scientists, resource managers, politicians, and humanitarians because the title implies a utopian goal of having the knowledge or capability to reconstruct past natural ecosystems or landscapes.

Watt (1973) noted that one could classify the variables (i.e., resources) that interact to govern ecological phenomenon into five major categories: matter, energy, space, time, and diversity. Interestingly it is the relationship of these variables (resources) that helps to define the above-mentioned fields of study. For example, landscape ecology focuses on the development of spatial heterogeneity at large temporal/spatial scales (Risser et al. 1984) and on how this heterogeneity is related to ecological phenomenon (Urban et al. 1987). Likewise, the concept of sustainability (Barrett 1989, 1992) and especially the Sustainable Biosphere Initiative (SBI; Lubchenco et al. 1991) prepared by the Ecological Society of America outline the need to address how the integration and management of these resources must be better understood if future generations are to enjoy a quality global environment. Our poor understanding of and inability to address the management of these resources in a transdisciplinary manner impedes the goals of restoration ecology as an integrative field of study, and is also a major reason why it is unlikely, if not impossible, to reconstruct or restore past landscapes.

World Resources

I will briefly describe each of these resource variables and how each has been either degraded, ignored, or mismanaged during this century. As they are all interrelated, these variables are not intended to be discussed in any ranked order.

Matter. I will focus here on only one major category of matter, namely, organic biological material (i.e., biomass) either provided from or supported by global net primary productivity (NPP). Because valid estimates of and limits to the Earth's primary productivity are known (Whittaker 1975), and because the present human population of 5.3 billion is adding 1.8 million persons per week to this global system, therefore it is likely that there will soon be too much biomass stored directly in or directed toward human systems (including the highly subsidized agricultural food system used to supply food for humankind) and not enough biomass made available for the diversity of other species that inhabit this planet.

Restoring natural systems so that former rates of net primary productivity are realized will require that soil nutrients (including the microorganisms necessary to cycle these nutrients) be restored to previous levels. It now seems unlikely that previous lithospheric deposits of phosphorus, for example, now deposited on the oceans' floors, can be recycled back to terrestrial systems in a cost–effective manner to restore such systems (the concept of net energy, to be discussed later, will defeat this strategy, especially for poor or developing nations).

Energy. Frequently when one speaks of restoring a system, especially an ecological system of large spatial scale, one tends to neglect the laws of thermodynamics or the concepts of net energy and energy quality. Net positive energy requires that energy yield (e.g., crop yield) must be greater than the energy subsidies necessary to sustain yield (Odum 1989). One must also note that the food production system in developed countries is a highly subsidized system (Barrett et al. 1990) in which high-quality energy (e.g., fossil fuels, pesticides, and commercial fertilizers) is used as subsidies to increase the yield of a diffuse, low energy quality, solar–based system.

Unfortunately, high-quality energy subsidies will be required to restore many of the degraded natural systems because of the previous loss of soil nutrients, biotic diversity, and food-chain regulatory processes. For this reason, society probably will not be able to restore past landscapes on a global scale. If energy were used as a worldwide "ecological currency" (rather than dollars, for example), then "ecological bookkeeping" would quickly illustrate why the concepts of net energy, energy quality, and energy subsidies will greatly impede the goals and strategies of the restoration process and of the invented landscape (see Turner, "The Invented Landscape," this volume), especially on a large temporal/spatial scale.

Time. One way to avoid the increased use of energy subsidies is to permit nature to "heal herself" (i.e., soil chemistry and biotic diversity will slowly be restored through natural developmental and coevolutionary regulatory processes—assuming, of course, that keystone species have not become extinct and are able to disperse into the system being "restored"). Time's value as a resource is often ignored, and, consequently, we continue to subsidize and manage such systems on a shorter and smaller temporal/spatial scale. When this occurs, usually "the rich get richer, and the poor get poorer." With the human population rapidly increasing within a defined spatial system—the Earth—the human-subsidized restoration process will eventually mean additional starvation, especially in underdeveloped countries, resulting in less chance of developing a global sustainable societal system.

One must also be cognizant that natural regulatory process and phenomenon have evolved on varying temporal scales. For example, one can observe annual growing seasons and reproductive cycles, four-year and ten-year mammalian population cycles, 350–400 year fire cycles in perturbation-dependent systems such as Yellowstone National Park, and glacial cycles of thousands, if not millions, of years. Time must be viewed in a hierarchical framework (O'Neill et al. 1986) when attempting to outline and understand restoration processes and management strategies.

Space. Space must also be viewed from a scaling (hierarchical) context. Restoring an abandoned farm in Sand County, Wisconsin (Leopold 1949) was both an interesting and intellectual project, but restoring millions of hectares of watersheds, abandoned strip-mine lands, or eroded croplands (not to mention the rate of deforestation in the tropical rain forest) poses a problem of scale never before encountered by humankind.

Such tasks in the field of restoration ecology will require unparalleled cooperation from both a national and international perspective. The coming century will probably demand greater problem-solving capabilities as the world's human population continues to increase and as problems related to world food production, energy resource management, biotic diversity, and environmental contamination continue to magnify (Barrett 1989; Lubchenco et al. 1991). Only by addressing these problems on a pragmatic temporal/spatial scale will meaningful resource management strategies and research approaches develop. Piecemeal, small-scale, nonintegrated approaches will likely lead to continued habitat and global destruction; holistic, large-scale, integrated approaches will likely result in the research, education, and environmental decision-making agenda (i.e., the Sustainable Biosphere Initiative) as outlined by the Ecological Society of America (Lubchenco et al. 1991) and the International Association of Ecology (Huntley et al. 1991).

Diversity. Diversity, especially biodiversity, is one of those ecological concepts that caught the nation's attention during the 1980s. A report to the National Science Board of the National Science Foundation in 1989 entitled *Loss of Biological Diversity: A Global Crisis Requiring International Solutions* captures the essence of biodiversity crisis and the scale of cooperation needed to achieve its resolution.

Human prosperity is based largely on the ability to utilize biological diversity. Society must understand the properties of plants, animals, fungi, and microorganisms as sources of food, clothing, shelter, and medicine. Unfortunately our understanding of science and engineering in the United States has regressed during the past two decades compared to other nations. We are facing a time of considerable challenge without the trained personnel to address these problems adequately or to take advantage of these research opportunities.

It is interesting to note, however, that the "biodiversity crisis" has led to the evolution of two major paradigms in the 1980s. Conservation biology has rapidly evolved as a new interdisciplinary field of study (Gibbons 1992). The Society of Conservation Biology now has more than five thousand members and its own journal (*Conservation Biology*) that focuses on biodiversity from several levels of ecological organization (genetic, species, ecosystem, landscape, and global).

Landscape ecology has also rapidly developed because of the need to understand the processes and dynamics of landscape patterns (largely the result of habitat fragmentation), the spatial and temporal interactions and exchanges of biotic and abiotic materials across the landscape, the influences of landscape patterns (i.e., spatial heterogeneity) on the spread of disturbance, and the management of the spatial heterogeneity for societal benefit and survival (Risser et al. 1984). A landscape is a mosaic of elements (patches and corridors) generated at various scales. We do not understand how best to manage this landscape mosaic in order to maximize genetic, species, niche, and habitat diversity. How do we restore biodiversity at several levels of integration simultaneously?

Several emerging paradigms (including restoration ecology) within the area of applied ecology require integrative concepts, approaches, and perspectives if they are to contribute to the resolution of societal and global problems in the future (see table 1). As noted by Chadwick (1990), "The most difficult and expensive way to pursue conservation would be to rebuild biological communities practically from scratch. Yet this emerging field of conservation, called restoration biology, is what may be required in some heavily populated and developed regions of the East." My prediction is that the concept of net energy, the laws of thermodynamics, the temporal/spatial scales involved (not to mention the necessary financial resources) will dictate that this utopian view of restoration ecology will give way to what I term sustainable ecology: a sustainable approach to the integration of matter, energy, time, space, and diversity resources. This new field of study will rely heavily on using natural ecosystems as model systems in the design of landscapes, including the increased use of solar-based energy, recycling of resources, and maintenance of biotic diversity for purposes such as pest management and community regulation (Barrett 1990). This approach will take us well into the twenty-first century because we first must revitalize our educational, research, political, socioeconomic, and service systems at most levels of organization.

The following case study and a list of six lessons we should have learned from the twentieth century outline an approach and recommendations to ensure that future generations not only survive, but manage world resources based on ecological knowledge rather than on regulatory mandates or large-scale human conflict.

Case Study

A relationship likely exists between the duration of a long-term perturbation and the length of time required to restore an ecosystem or landscape that has been degraded or simplified. As one piece of evidence for this statement, we investigated the effects of applying contrasting types

Table 1. Summary of emerging paradigms, major integrative concepts, and interdisciplinary approaches needed to revitalize our educational, research, and industrial institutions as we enter the twenty-first century.

Emerging Paradigms
    Restoration ecology
    Conservation biology
    Disturbance ecology
    Agroecology
    Landscape ecology
    Ecological toxicology

Integrative Concepts and Theory
    Hierarchy theory
        Temporal/spatial scale
        Levels of ecological organization
        Trophic levels and dynamics
    Sustainability
        Net energy
        Optimum landscape connectivity
        Global climate change
    Ecological diversity
        Genetic
        Species
        Niche
        Landscape
        Ethnic

Interdisciplinary Approaches and Technologies
    Scientific method
    Cost-benefit analysis
    Cybernetics
    Problem-solving algorithms
    Modeling
    Geographic Information Systems (GIS) and technologies

of nutrient enrichment (municipal sludge or fertilizer) on old-field eco-systems for eleven years (1978-88) following agricultural practice (for details, see Anderson and Barrett 1982; Maly and Barrett 1984; Carson and Barrett 1988; Levine et al. 1989; Bollinger et al. 1991). This long-term investigation was conducted at the Miami University Ecology Research Center located near Oxford, Ohio. The old-field was divided into eight 0.1-ha plots: three plots received monthly (May–September) applications of sludge (Milorganite; 6-2-0, N-P-K), three plots received an equivalent nutrient subsidy of fertilizer, and two plots were left as untreated controls.

Plant species richness was significantly higher in control plots than nutrient-enriched plots from the second year of secondary succession thereafter. Nutrient-enriched plots also had significantly higher rates of annual net primary productivity than control plots during the later stages of ecosystem development. Summer annuals, especially *Ambrosia trifida, A. artemisiifolia*, and *Setaria faberii*, dominated nutrient-enriched plots throughout the study, whereas winter annuals and biennials replaced summer annuals in control plots during years 2-3, followed by a series of perennials (e.g., *Aster pilosus, Poa compressa*, and *Solidago canadensis*) during years 4-11.

We are now attempting to restore these modified nutrient-enriched plots to control levels concerning both structural (e.g., plant species richness) and functional (e.g., annual net primary productivity) ecosystem parameters. Each of the former long-term (eleven-year) nutrient-enriched plots (sludge and fertilizer) were subdivided into four subplots (tilled, limed, tilled and limed, and unmanipulated). Subplots in control plots were either unmanipulated or tilled. These subplots were tilled and/or limed to investigate biological (e.g., seed bank disturbance) or chemical (e.g., soil pH) mechanisms of ecosystem recovery.

Differences in plant community structure three years following these manipulations reflect long-term treatments rather than short-term manipulations. For example, control plots remain dominated by perennials such as *Solidago canadensis*. Although former fertilizer plots remain dominated by the summer annual *Ambrosia trifida*, perennials such as *Aster pilosus* increased significantly indicative of secondary succession (i.e., of community restoration). Sludge plots remain dominated by the summer annuals, *A. trifida, Chenopodium album*, and *Polygonum pennsylvanicus*. Increased soil nitrate levels in former sludge-treated plots probably account for these differences between nutrient-enriched plots. Thus, despite the restoration of soil pH values to control levels, it appears that it will take several years to restore former nutrient-enriched systems to control levels regarding plant community composition and diversity. Further, active bacterial biomass and rates of litter decomposition (i.e., attributes of ecosystem function) remain greater in control than nutrient-enriched plots. Nitrate availability seems to have affected bacterial activity and nutrient cycling, which, in turn, controlled plant community structure and ecosystem development.

As mentioned earlier, there is an urgent need for long-term research (Callahan 1984), especially in the area of resource management (Barrett 1985). This case study provides evidence that a perturbed system will need many years to recover, including human intervention, following the application of long-term (chronic) nutrient subsidies. Perhaps even a decade will be required before attributes such as soil microbial composition and metabolic activities will be restored to control levels (Sutton et al. 1991). Such research must be addressed in an interdisciplinary manner and conducted on a long-term basis. Any restoration effort must be complemented by natural, solar-based processes and mechanisms if a sustainable system or landscape is the long-term goal. This perspective is also necessary if "ecosystem construction" (Jordan, " 'Sunflower Forest' ") is to contribute toward a healthy and viable relationship between solving societal problems and managing nature's resources.

Lessons Yet to Be Learned

As with the world resources described previously, these "lessons" or recommendations are not in prioritized order, nor are they intended to cover the human landscape. Rather, they represent current educational, administrative, and economic constraints that must be addressed if a field of study such as restoration ecology is to become more than a passing topic of interest and, indeed, underpin academic, governmental, and industrial structures and management processes during the decade and century ahead.

1. Establish rigorous inter- and transdisciplinary educational programs, including a fair and appropriate reward system. Despite the "Decade of the Environment" in the 1970s and the numerous centers, institutes, and programs dealing with environmental science, environmental education, and ecology established throughout the country and around the world, there remains the lack of a firm commitment by institutions of higher learning to initiate quality academic programs in areas such as restoration ecology, conservation biology, or landscape ecology. Administrators still tend to fund and reward the more traditional disciplinary programs, even to the extent that new "cutting edge" field of study must be taught as an "overload" or without valid joint appointments. Although disciplinary boundaries are dissolving everywhere, universities add to their problems by clinging to antiquated disciplinary structures and failing to adapt to new developments. "Academia is back in the twelfth century in terms of organization," notes physicist Donald Shapero of the National Academy of Sciences (Holden 1991). Unless universities truly become creative, including the establishment of a fair and equitable reward system for young investigators who desire to explore questions of an interdisciplinary nature, then restoration ecology is destined not to become a productive and enriching academic field of study for decades.

2. *Develop the concept of sustainability as an integrative approach for the management of natural, subsidized, and socioeconomic systems.* Barrett (1989) noted that the concept of sustainability appears to be not only a commonality, but perhaps the driving force behind several new emerging paradigms, including restoration ecology. The verb sustain is defined "to supply with necessities or nourishment to prevent from falling below a given threshold of health or vitality." Therefore, the unifying theme of these fields of study is frequently the implementation of lower-input sustainable agriculture, the maintenance of biotic diversity, and the restoration of a degraded environment. Unfortunately, we have not learned to use natural systems as a solar-powered model, including appreciating the complexity of its evolutionary regulatory mechanisms, when attempting to restore previously human-subsidized (e.g., agricultural or urban) systems on a sustainable basis.

3. *Conduct long-term landscape and global-level (holistic) research in an integrative manner.* The need for and importance of long-term ecological research at the ecosystem and landscape levels have been well addressed (Callahan 1984; Strayer et al. 1986; Likens 1989). Indeed, a network of Long-Term Ecological Research (LTER) sites funded by the National Science Foundation has now been in operation for over a decade. Although investigators working at these sites recognize the ecosystem and watershed concepts as units necessary to integrate processes (e.g., nutrient fluxes, community regulatory mechanisms, and energetics) within and between landscape elements, there still remains a paucity of information regarding how socioeconomic factors influence these processes or how best to integrate such parameters into long-term landscape-level investigations. These data are vital before resource managers are able to

manage and restore perturbed systems meaningfully. Perhaps an NSF LTER site focused on landscape or restoration ecology, including the human dimension (e.g., farmers and/or a representative town community) should be funded to address this need.

4. *Establish an ecological currency regarding the development of a long-term management perspective.* There is a dire need to establish an ecological currency (e.g., energy) rather than a strictly monetary currency regarding not only the value of managed landscapes, but also how to compare the costs and benefits of the restoration process. Dollars, energy, and environmental quality (E.Q.) units have in the past been used for impact assessment and in the development of a problem-solving algorithm (Odum 1977; Barrett 1985; Barrett and Bohlen 1991). I suggest that net energy be considered as the common denominator or currency for detailed analyses in the field of restoration ecology.

When energy is not employed, then it is imperative that "monetary credit" be fixed to such natural, solar-based ecosystem-level and landscape-level processes as water purification, nutrient recycling, prevention of soil erosion, primary productivity, wildlife habitat maintenance, and insect pest control, to name only a few. One must again be reminded that natural ecosystems are unsubsidized, solar-powered systems (Barrett 1990). Although human resources, fossil fuel, and nutrient subsidies will probably be necessary in the short term to restore perturbed landscapes, it is likely that these systems, if managed properly, will not require these increased subsidies on a long-term basis.

5. *Interface the academic-industrial-governmental triumvirate for meaningful and cost-effective educational, research, and service initiatives in the field of restoration ecology.* Restoring human-dominated landscapes (e.g., agricultural, strip-mined, or energy production sites) will require a renewed cooperation and trust between academia, industry, and government in order that those residing in the area to be restored benefit (intellectually, aesthetically, and monetarily) by the restoration process. At present all three entities have failed to integrate training, career opportunities, and policies in a sustained organizational framework and on a large-scale basis.

Industry urgently needs to provide a broader array of "real world" training connections with academia that link basic and applied components of the educational process (e.g., college-preparatory courses, undergraduate and graduate internships, worker retraining and career courses, and administrative seminars). Workers should have more say in the decision-making process, and universities need to provide a greater forum for problem-solving skills, conflict resolution, and career training objectives necessary for the twenty-first century.

As a nation we must address (indeed, admit) just how poorly our students are educated not only in the basic science and engineering fields, but also in the intellectual tools that are essential to integrate holistic concepts in order to solve problems and design studies in new fields such as restoration or sustainable ecology. Although numerous reports have either articulated this educational concern (such as, the "America 2000" report; Bush 1991) or attempted to outline a research agenda for the future (e.g., the Sustainable Biosphere Initiative; Lubchenco et al. 1991), few major cooperative initiatives have been developed to address this challenge.

6. *Each citizen should address societal topics and problems in a holistic context, rather than via a selfish or reductionist approach.* The field of restoration ecology provides the public with the opportunity to solve societal problems in a new and creative mode. For example, restoration ecology encompasses such topics and challenges as maintaining and restoring biodiversity (Chadwick 1990); the restoration of landscape linkages (Barrett and Bohlen 1991); the implementation of alternative agriculture on a regional and national basis (National Research Council 1989); the reduction of environmental contamination; and how best to wed humankind and nature in order that our human value and socioeconomic systems are changed on a large temporal/spatial scale. This requires that society change from the "me generation" to the "we generation"; that the restored system (society) benefit, rather than only the individual; and that social justice be viewed, understood, and restored regarding its relationship to the natural ecosystem and global processes on which society depends.

Conclusions

We, as a nation, will not be able to restore large-scale systems without simultaneously addressing societal problems such as human population growth, world food production, global climate change, tropical deforestation, and environmental contamination. At present, the youth of the world largely fail to recognize the commonalities and relationships between biological, physical, and social sciences and issues. For example, the laws of thermodynamics, the concept of carrying capacity, or the importance of diversity (social, genetic, or landscape) represent laws or concepts that can best be addressed in an integrative context. If new fields of study, such as restoration ecology, are to become vital areas of study during the coming decade, then our educational, research, and service organizational structures must change or adapt to meet these needs.

The old adage—"One must first identify a problem before one can hope to solve it" applies to society's current dilemma. For example, American students scored twelfth out of twelve countries and provinces on an international test conducted by the Educational Testing Service in 1988 (Perry 1991). Yet when asked how good they thought they were in math, 68 percent of the U.S. students replied "very good." Thus, we need to be objective and do much more than "preserve" what is already in place regarding our educational goals for the future. We need to recognize and restore learning, including a better understanding of the integration of knowledge in the learning process, if educational standards are to improve on a national basis.

An objective of understanding, restoring, and managing a landscape unit (e.g., a watershed) in a sustainable manner (including farms, small towns, and landscape corridors) could serve as a model project to address these integrative needs. At the very least such a project would focus attention on integrating social, biological, physical, and economic factors at a scale of resolution that could be used as a model for addressing problems nationally. Perhaps success at the landscape level would facilitate the establishment (or at least the coordination) of a federal unit of research and education, like a National Institute for the Environment, in which such agencies as the National Science Foundation, the Environmental Protection Agency, the U.S. Department of Agriculture, and the Department of Energy, among others, would focus on projects that wed basic and applied science, unite disciplinary and interdisciplinary approaches, encompass educational concepts (e.g., hierarchy theory, net energy, and sustainable resource management), and include the public efficiently and cost-effectively.

Our present situation, which includes increased budget deficits, compartmentalized education, and piecemeal research, is not structured to address and solve societal problems. Let's discard the early-twentieth-century educational and research agenda in favor of a twenty-first-century cooperative, integrative, and problem-focused approach to societal issues and values. We have already mortgaged the economic future of our children; let's not continue to degrade and neglect their natural and human resources as well. A long-term and large-scale integrative program for landscape restoration provides a national agenda to adopt the suggested recommendations and to begin managing our natural resources in a sustainable manner.

REFERENCES

Aber, J. D., and W. R. Jordan III. 1985. "Restoration Ecology: An Environmental Middle Ground." *BioScience* 35 (July/August): 399.

Anderson, T. J., and G. W. Barrett. 1982. "Effects of Dried Sewage Sludge on Meadow Vole (Microtus pennsylvanicus) Populations in Two Grassland Communities." *Journal of Applied Ecology* 19 (December): 759-72.

Barrett, G. W. 1984. "Applied Ecology: An Integrative Paradigm for the 1980s." *Environmental Conservation* 11 (Winter): 319-22.

_____. 1985. "A Problem-solving Approach to Resource Management." *BioScience* 35 (July/August): 423-27.

_____. 1989. "A Sustainable Society." *BioScience* 39 (December): 754.

_____. 1990. "Nature's Model." *Earthwatch* 9 (April): 24-25.

_____. 1992. "Landscape Ecology: Designing Sustainable Agricultural Landscapes." *Journal of Sustainable Agriculture* 2 (3): 83-103.

Barrett, G.W., and P. J. Bohlen. 1991. "Landscape Ecology." In W. E. Hudson, ed., *Landscape Linkages and Biodiversity*. Washington, D.C.: Defenders of Wildlife. 149-61.

Barrett, G.W., N. Rodenhouse, and P. J. Bohlen. 1990. "Role of Sustainable Agriculture in Rural Landscapes." In C. A. Edwards et al., eds., *Sustainable Agricultural Systems*. Ankeny, Iowa: Soil and Water Conservation Society. 624-36.

Bollinger, E. K., S. J. Harper, and G. W. Barrett. 1991. "Effects of Seasonal Drought on Old-field Plant Communities." *American Midland Naturalist* 125 (January): 114-25.

Bush, G. 1991. "America 2000." *Congressional Digest* 70 (December): 294-95.

Callahan, J. T. 1984. "Long-term Ecological Research." *BioScience* 34 (June): 363-67.

Carson, W. P., and G. W. Barrett. 1988. "Succession in Old-field Plant Communities: Effects of Contrasting Types of Nutrient Enrichment." *Ecology* 69 (August): 984-94.

Chadwick, D. H. 1990. "The Biodiversity Challenge." In W. E. Hudson, ed., *Landscape Linkages and Biodiversity*. Washington, D.C.: Defenders of Wildlife. 1-11.

Gibbons, A. 1992. "Conservation Biology in the Fast Lane." *Science* 225 (January): 20-22.

Holden, C. 1991. "Career Trends for the '90s." *Science* 252 (May): 1110-20.

Huntley, B. J., E. Ezcurra, E. R. Fuentes, K. Fujii, P. J. Grubb, W. Haber, J. R. E. Harger, M. M. Holland, S. A. Levin, J. Lubchenco, H. A. Mooney, V. Neronov, I. Noble, H. R. Pulliam, P. S. Ramakrishnan, P. G. Risser, O. Sala, J. Sarukhan, and W. G. Sombroek. 1991. "A Sustainable Biosphere: The Global Imperative." *Ecology International* 20: 1-14.

Jordan, W. R., III, M. E. Gilpin, and J. D. Aber, eds. 1987. *Restoration Ecology: A Synthetic Approach to Ecological Research*. Cambridge: Cambridge University Press.

Leopold, A. 1949. *A Sand County Almanac*. New York: Oxford University Press.

Levine, M. B., A. T. Hall, G. W. Barrett, and D. H. Taylor. 1989. "Heavy Metal Concentrations during Ten Years of Sludge Treatment to an Old-field Community." *Journal of Environmental Quality* 18 (October): 411-18.

Likens, G. E. 1989. *Long-term Studies in Ecology: Approaches and Alternatives*. New York: Springer-Verlag.

Lubchenco, J., A. M. Olson, L. B. Brubaker, S. R. Carpenter, M. M. Holland, S. P. Hubbell, S. A. Levin, J. A. MacMahon, P. A. Matson, J. M. Melillo, H. A. Mooney, C. H.

Peterson, H. R. Pulliam, L. A. Real, P. J. Regal, and P. G. Risser. 1991. "The Sustainable Biosphere Initiative: An Ecological Research Agenda." *Ecology* 72 (April): 371–412.

Maly, M. S., and G. W. Barrett. 1984. "Effects of Two Types of Nutrient Enrichment on the Structure and Function of Contrasting Old-field Communities." *American Midland Naturalist* 111 (April): 342–57.

National Research Council. 1989. *Alternative Agriculture*. Washington, D.C.: National Academy Press.

National Science Foundation. 1989. *Loss of Biological Diversity: A Global Crisis Requiring International Solutions*. Report to the National Science Board. Washington, D.C.

Odum, E. P. 1977. "The Emergence of Ecology as a New Integrative Discipline." *Science* 195 (March): 1289–93.

_____. 1989. *Ecology and Our Endangered Life-support Systems*. Sunderland, Mass.: Sinauer Associates, Inc.

O'Neill, R. V., D. L. DeAngelis, J. B. Waide, and T. F. H. Allen. 1986. *A Hierarchical Concept of Ecosystems*. Monograph of Population Biology 23. Princeton: Princeton University Press.

Perry, N. J. 1991. "Where We Go from Here." *Fortune* 124 (October): 114–29.

Risser, P. G., J. R. Karr, and R. T. T. Forman. 1984. *Landscape Ecology: Directions and Approaches*. Special Publication No. 2. Champaign, Ill.: Illinois Natural History Survey.

Strayer, D., J. S. Glitzenstein, C. G. Jones, J. Kolasa, G. E. Likens, M. J. McDonnell, G. G. Parker, and S. T. A. Pickett. 1986. *Long-term Ecological Studies: An Illustrated Account of Their Design, Operation, and Importance to Ecology*. Occasional Publication No. 2. Millbrook, N.Y.: Institute of Ecosystem Studies, New York Botanical Garden.

Sutton, S. D., G. W. Barrett, and D. H. Taylor. 1991. "Microbial Metabolic Activities in Soils of Old-field Communities Following Eleven Years of Nutrient Enrichment." *Environmental Pollution* 72 (1): 1–10.

Urban, D. L., R. V. O'Neill, and H. H. Shugart, Jr. 1987. "Landscape Ecology: A Hierarchical Perspective Can Help Scientists Understand Spatial Patterns." *BioScience* 37 (February): 119–27.

Watt, K. E. F. 1973. *Principles of Environmental Science*. New York: McGraw-Hill.

Whittaker, R. H. 1975. *Communities and Ecosystems*. 2d ed. New York: MacMillan.

Art and Insight in Remnant Native Ecosystems

Orie L. Loucks

Like many others, I have sought out museums and art galleries housing the originals of well-known paintings or sculptures. Whether in Greece or China or France, viewing such works remains a clear and vivid memory for me, when much else from those visits is blurred or forgotten. In this essay I will argue that seeing and comprehending unique remnants of native ecosystems is a learning experience about a unique art form, and one to be cherished. Thus, in contrast to arguments made by William Jordan and Frederick Turner, the highest priority must be attached to preserving the species and ecosystems that remain. Seeing and understanding restorations can be important, obviously so when few originals are available in any form, but still they are copies and less instructive than originals. A restoration makes the same impression as a new manufacturing plant: wonderful assemblages of nuts and bolts, put together like clockwork, a feat of modern technology, but lacking spirit, place, or lasting significance.

One truth is that the remaining natural systems dispersed across our landscapes show the subtle effects of pesticides, air pollution, exotic species, and the absence of fire. Virtually no site is truly an "original." Some level of restoration is a sine qua non for the preservation of natural systems, as it is for great works of art. I draw a distinction, however, between the maintenance restoration required at sites where the remaining natural history is of local or regional significance, and undertakings to "restore" a community or ecosystem in places where most of the species and ecosystem functions either never occurred or have been lost for decades. Such restoration (or reproduction, in Turner's terms) is feasible, of course, over very long periods of time, but with a likelihood that external maintenance will be required for decades if not centuries.

To develop my point that natural systems (even with some maintenance) are jewels to be cherished, and restorations are simply engineered systems, marvels of our new nature technologies, I will summarize experience from two important sites. One is a wetland complex in Switzerland threatened in the 1960s by an airport runway extension, and the other is a small mountain preserve along the Pearl River in southern China, a site surviving high human densities and centuries of exploitation in the surrounding landscape.

Transplanting the Zurich Airport Wetland

One of the most interesting feats of restoration (engineering), or nature reproduction, is the program begun in 1968 in Switzerland to relocate a sedge-cottongrass, meadow, and fen complex threatened by a runway extension for the Zurich airport. I first heard of the physical moving and replanting of this wetland to a new site three years later, and have sought to follow its progress over more than twenty years, as I had followed the University of Wisconsin Arboretum restorations before that. Professor Dr. Frank Klötzli of the Zurich Technical University participated most closely in the monitoring of the project and has published extensively on its subsequent progress (1975, 1981, 1987). Moving an entire ecosystem complex intrigued me: I wondered whether everything that is essential in a natural system might possibly be saved and re-created by such measures, if the moving could be engineered carefully. Perhaps this would be the solution to the difficult worldwide problem of saving threatened ecosystems from suburban and industrial development!

Klötzli describes the Zurich airport site as an undulating alluvial plain, the bed of a postglacial shallow lake dotted by low morainic hills. Because of this physical variability, a diversity of wet and relatively dry sites is found, many of which have been modified further by selective mowing regimes (Klötzli 1987). The result is a mosaic of open meadows and peaty depressions, interspersed by seminatural woodlands. Species diversity is high, with five hundred species in an area of 10 square kilometers; several of these species are viewed as rare or very rare (Klötzli 1969, 1975).

A program of mass transplanting of at least the most important wetlands threatened by the runway was begun in 1968, sponsored and paid for by the airport and the Department of Public Construction of the Zurich Canton. Before undertaking such transplantings, Klötzli notes that key physical factors needed to be evaluated, and conditions at the new site had to be matched to those at the old site. Water and nutrient variables were the most important. For water, the mean depth to water within the peat profile, the seasonal fluctuations of the groundwater, and the duration of inundations in the root horizon had to be matched or managed carefully. For nutrients, concentration of available nitrogen and phosphorus within the rooting zone, the profile pH, and exchangeable cations in the peat had to be matched.

A major problem in mass transplanting relates to the weight of the peat or soil and sod layer to be transported. The blocks to be moved had to be as large as possible to maintain homogeneity of the transplanted community in the new site. Generally, the turf blocks were transported on pallets of about 90 × 130 cm in size. Considering the depth through the whole rooting zone (30-50 cm deep), weights ranged up to 3/4 to 1 metric ton, so the machinery had to be heavily built. Certain types of front-end loader equipment were found to be satisfactory for lifting the block onto a pallet, and the pallets onto vehicles (Klötzli 1987). To move the blocks to the new site, trucks of 5-10 ton capacity were used. Then the blocks had to be unloaded at the new site, using forklift elevators to move them from the truck to the placement site. Finally the turf blocks were slid from the pallets, usually with ropes (see fig. 1). The blocks had to be reset precisely, and in the same block pattern from which they had been dug up. In the most difficult cases, placement was accomplished by hand, using ropes and eight to ten men per pallet. Getting a uniform surface during the resetting of the turf proved to be a problem; a surface free of cracks or fissures was especially hard to obtain. Tilting up of the borders at the edge of blocks because of their being placed too closely was found to change aeration along the border, stimulating mineralization processes (fig. 1). For many reasons the total surface area moved had to be kept small: normally only 100 square meters were moved for any one community, although for the wettest sites 200 to 5,000 square meters were moved and reset.

At the end of the first growing season, and for the years following, plant occurrences and distributions on the new surfaces have been monitored and analyzed, with emphasis on the presence and vitality of key

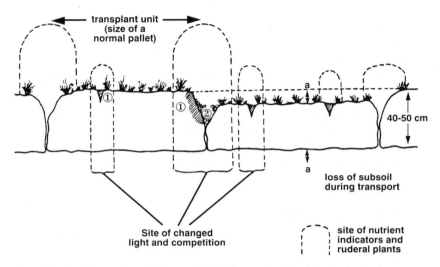

Figure 1. An illustration of the method of transplanting blocks from the wetland complex, showing resulting changes in sward structure, drainage and mineralization processes, and competition. The cavities (1) lead to mineralization of humus; the fissures (2) lead to mixing of humus and subsoil. (Adapted from Klötzli 1980.)

indicators, e.g., dryness, disturbance, or N–indicators (Klötzli 1987). For a few years, weeds tended to invade the cracks between transplant units. Most of the communities that were moved suffered "transplantation shock," in which some species increased, others disappeared (or fluctuated strongly), and still others invaded. These changes were mostly due to the disturbance during transplanting (Klötzli 1987). The species that disappeared almost completely included representatives of the Orchidaceae, Apiaceae, and *Equisetum*.

In summary, Klötzli (1987) notes most changes were due to disturbances that affected the availability of water or nutrients, influenced in part by the influx of nutrients from differences encountered in the surrounding areas. Another factor that may have induced change was the closer proximity of contrasting communities in the transplanted landscape, leading to in–migration of new species. Thus, he says, mass transplantings may be applicable only for rare plant communities threatened with obliteration. The costs are very high, but the probability of success can be reasonably high if the sites are carefully matched and subsequent management is appropriate and sustained.

Shortly after hearing of the apparent feasibility of moving and reconstructing the wetlands around the Zurich airport, I was confronted with a similar proposal by the U.S. Corps of Engineers in Wisconsin. They were planning to build a dam and reservoir on the Kickapoo River and proposed to save an endangered primrose (*Primula misstasinica*) on the sandstone cliffs along the river by physically removing the sandstone blocks on which the flower grew and transporting them to be installed in other cliffs in southwestern Wisconsin. Money was not a problem, at least then, but trying to guess what cliff faces or valleys might support a primrose population for future generations clearly was problematical. It seemed that our ingenuity and success in Europe was leading only to a new type of transplant garden, with all the maintenance that entails. The dam was never finished.

For the Swiss example, reproduction of the system at a new site should have been simpler than complete rebuilding of the system, species by species. Klötzli (1987) concludes that mass transplanting can be considered for managed conservation of vegetation, but not as a substitute for conservation in situ. He views it as an expensive solution appropriate only when other forms of conservation cannot be implemented. "A transplanted and reconstituted or renaturalized surface should not be considered as good as a naturally grown and original surface," he says. "A good copy of the Mona Lisa is still not *the* Mona Lisa" (96).

A Biosphere Reserve in Southeast China

Ding-hu Shan (Ding-hu mountain) is a protected area comprised of a few low mountains on the floodplain of the Pearl River west of Guangzhou, China. About fifty million people live within a three-hour travel radius (about 200 km by train), and a large number of these people seem to know of and appreciate the preserve. As many as thirty thousand visit the site on religious holidays.

Ding-hu Shan has been accorded long-term protection for religious reasons. It is a sacred grove where the Buddha's spirit lives, on occasion. But whence derives the respect accorded this site a thousand or so years ago, when it was probably first given special treatment? Is the source of attention any different from that addressed to nature preserves in eastern Europe that are established now on what were formerly the "king's hunting lands"? We will never know. All such places are special now, as they probably were then, because of their biological uniqueness. All have been interwoven with the cultural fabric of their countries for centuries. It appears that only in the industrialized nations, the progressive "West," are such places judged to have no art or spiritual or cultural significance, only scientific and educational value. Turner and Jordan now propose that in the West we accord unique aesthetic significance to the reengineered sites (restorations) because natural systems are too threatened to protect any longer. It seems that only in the West could we confuse the role of an original with that of a restoration, or elevate the restoration of nature to the equivalent of nature's originals.

One arrives at Ding-hu Shan by train, bus, or car from Guangzhou (Canton), the busy metropolis, port city, and capital of Guangdong Province. Guangzhou is the inland counterpart of Hong Kong, and, with economic reform, has become the center of the economic miracle of China. The road route traverses a cross section of county towns making up the dispersed suburbs of Guangzhou, followed by a few tens of miles of valley-bottom rice paddies, alternating at high frequency with rural villages in the Pearl River plain. Low hills, where they occur, have been overexploited and support an irregular cover of weeds, a few flowering shrubs, and a scattering of Masson pines, the latter planted. The hills rise steeply as we approach the town of Zhaoqing, and the closed forest cover is mostly a mixture of pine and eucalyptus, an import from Australia. My colleagues tell me these are "restorations"; before the revolution all these hills had been laid bare. The pine and eucalyptus are planted because they are able to survive in the harsh-disturbance environments. The litterfall in the forest is now gathered once or twice a year to supplement cooking fuels.

One wonders what were the natural systems that once populated this landscape, this mosaic of floodplain, low hills, and the steep mountainsides. For how many millennia has it all been gone? How many species have been lost directly from the physical transformation? How many were lost indirectly through the loss of pollinators or dispersal agents?

At Zhaoqing we turn to the north and almost immediately see the scientific headquarters of the Biosphere Reserve, so designated by UNESCO in 1980. Research has been under way since the site was first established as a nature reserve in 1956. Another two or three kilometers up a mountain road and one is in the ancient Temple Forest. Surveys during the past thirty years show the preserve as a whole (some 1,155 hectares) supports 1,740 species of plants, almost equal the flora of the entire state of Wisconsin. About one-fifth of the species are introductions, mostly accidents of centuries of casual human activity, but many resulted from recent programs of "restoration" on lands around the periphery of the central protected area. The modest list of five hundred insect species known to date indicates this group is only beginning to be investigated.

Fundamentally, the centuries-old Buddhist temple attracts most of the public. Spaces for prayers are occupied on weekends and holidays, and wooden splints with incense lighted and burning are placed everywhere the Buddha might be: at statues, at portals, and especially in the vicinity of the most ancient trees. The immense variety of fauna and flora suggests a system of great integrity, something one admires instinctively, and one wonders if the Buddha came knowing all this, or did he learn it as we do.

Altogether there are 320 species of timber trees in the preserve. Research shows that about ten of these should be considered threatened: either the seeds cannot be germinated, or seedlings have not been seen for thirty years. *Erythrophlem fordii, Tsoongiodendron odorum,* and some Podocarps are among those threatened. Research is being focused on the reproductive biology of the species, but the state does not consider this a high priority. What does it matter if a pollinator has been lost? Perhaps these species are reproducing somewhere else, although no one is sure where that would be. No one is sure which of the understory orchids or ferns seen along the trail (including tree ferns) should be thought of as endangered.

The preserve is made up of a cluster of low mountains, the highest point rising to 1,000 meters. A profound zonation exists in the montane vegetation because winter freezing is severe at the highest elevations, and rare on the valley floor. Low rhododendron forests dominate the highest elevations. The area is large enough to support many local to wide-ranging birds and mammals, although these populations are hard to monitor. On one occasion a tiger (apparently a visitor from another remote area to the northwest) had just been seen on a trail I was to take, but rain precluded our expedition into that area. The preserve director speculated that the tiger might be too valuable for aphrodisiac and other commercial purposes to last very long at Ding-hu Shan, the staff of wardens notwithstanding.

In spite of the challenges to modern protection at this site, physically and biologically, one becomes totally engaged by the significance of what is being protected. It is both a scientific and a spiritual wonder. The integration of climate, topography, plants, animals, and other biota, through evolution, into a single, sustaining system is a feat of nature that no one can contemplate as a restoration. The experience is humbling in the most constructive sense; without needing quantitative measures of the system, one is afforded insight about great masters and about the importance of originals.

Discussion

Do these two examples satisfactorily represent the fundamental difference between preserving the remnants we have of native ecosystems, or creating restorations of them? Is protection of such art forms important, or is knowing and learning about the reproduction of natural systems the art and the priority? Obviously, it is through protection that we have reference cases, the systems we imitate in restorations. The Zurich airport example almost qualifies as preservation of an "original," but in the end it is not a system wrought by the natural processes that occupy the new site, and there is no longer an indigenous site. It is modern engineering at its best (or worst), a novel approach to the creation of transplant gardens, differing only in the method of planting from those we have pursued since the Middle Ages.

What of the protected sites? Are they all destined to fail from being too small in relation to species' breeding requirements or the magnitude of threats and insults imposed on nature by distant technologies? Some will fail, but I have worked on the design and implementation of nature preserves since 1955 (at that time for the Canadian government), and mistakes or failures during these forty years are overwhelmed by the successes. Not only has time healed damage at sites that were threatened years ago, but humans have learned more about what the species need from a supporting landscape, and about how to provide it. This process requires a society that values the opportunity to understand complex systems whose secrets are revealed only slowly, as at Ding-hu Shan and other areas, rather than a society that believes it knows enough now to create all the systems we need. There will be no secrets in these newly engineered sites; all will have been specified.

A few years ago, I wrote about the resilience of the forests of the Great Lakes region (Loucks 1983). They withstand dislocations by fire and storms over periods of decades to centuries; they change because of climatic cycles over millennia, and have recovered from devastation of glacial advances only ten thousand years ago. Although human activity has added yet another threat to forests, and restoration seems to mitigate that threat locally, the real resilience of forests is evident in the repeated long-term recovery. No human restorations have been sustained satisfactorily for the life span of even one rotation of a forest (three hundred years or so). But as is shown by these forests' recovery from glaciation, human guidance is not necessary—only respect and modest protection are. Recovery of natural systems after this era of human intervention is almost certain; the apparent concern of humans in restoring the forests now is a concern of society for resources to sustain itself, not for the forest itself. Having remnants of native ecosystems, rather than restorations, is critical to understanding how the natural recovery will take place when *Homo sapiens* is no longer the agent of destruction.

Turner and Jordan have proposed that we celebrate the human potential to restore natural systems. We are asked to view such restoration as an outlet for our creative genius, or, if we get the species interactions right, as a technological tour de force. I think we can all celebrate the proclivity of primates to imitate, and to be creative in the process, but this cannot be extended to ecosystems. Copies made by chimpanzees entertain us, partly because we are amused by how little the chimp understands of his creation. We should consider seriously how little we know or understand of our own imitations when we propose to restore whole ecosystems. Looking at the global panoply of risks to resources, the level of human ignorance is most evident in the destruction we have wrought, not particularly in how gross the restorations are. Originals, small and isolated though they are, remain our only reference point for understanding the earth.

REFERENCES

Klötzli, F. 1969. "Die Grundwasserbeziehungen der Streu und Moorwiesen im noördlichen Schweizer Mittelland." *Beitr. Geobot. Landesaufn* 52. 296 pp.

———. 1975. "Naturschutz im Flughafengebiet—Konflikte und Symbiose." *Flughafen-Information* 3: 3-13, 21.

———. 1981. "Zur Reaktion verpflanzter Oekosysteme der Feuchtgebiete." *Dat. u. Dokum. Umweltschutz* 31: 107-17. Stuttgart: Univ. Hohenheim.

———. 1987. "Disturbance in Transplanted Grasslands and Wetlands." In J. van Andel et al., eds., *Disturbance in Grasslands*. Dordrecht: Dr. W. Junk Publishers. 79-96.

Loucks, Orie L. 1983. "New Light on the Changing Forest." In Susan L. Flader, ed., *The Great Lakes Forest—An Environmental and Social History*. Minneapolis: University of Minnesota Press. 17-32.

Natural Forest Management of Tropical Rain Forests: What Will Be the "Nature" of the Managed Forest?

David L. Gorchov

In recent years the mass media have joined biologists in decrying the loss of tropical rain forests (TRFs). The value of these forests and the costs of deforestation are now well known. Tropical rain forests contain at least half of the world's plant and animal species. Many environmentalists argue that each species has intrinsic value and the right to exist, though few declare that all species are of equal value—despite Frederick Turner's claim that this extreme position is a principle of the "ecological religion" (see Turner, "The Invented Landscape," this volume). But even with the human-centered approach that Turner advises, we have cause to be alarmed about extinctions of rain forest species.

Tropical plants are particularly likely to contain medically valuable compounds, but the vast majority have not yet been investigated (Soejarto and Farnsworth 1989). Among those that have already proven of great medical value are *Cinchona*, which provides quinine against malaria, and Rosy Periwinkle, which provides vincristine and vinblastine against Hodgkin's disease and childhood leukemia (Soejarto and Farnsworth 1989; Miller and Tangley 1991).

The plants that give us rubber, palm oil, cane sugar, citrus fruits, bananas, coconut, and chocolate all originated in TRFs (Whitmore 1990). In addition, several wild tropical forest species have the potential to yield important new food or industrial products (Plotkin 1988). Crossbreeding with wild relatives has been invaluable in introducing desirable traits such as disease resistance into numerous crops; among the crop species that have potentially important wild tropical forest relatives are rice

(Vaughan and Sitch 1991), coffee, potatoes, and avocado (Plotkin 1988). Even wild species that are unrelated to any domestic species may provide desirable traits using the rapidly developing techniques of genetic engineering (Myers 1983).

Some of the world's remaining indigenous cultures live in TRFs. These peoples depend on the forest for food, construction materials, medicine, and so on, and their knowledge of the uses of the forest species is proving invaluable in the search for plants of value to the outside world (Plotkin 1988; Posey 1990). Many indigenous peoples have developed agricultural systems that are demonstrating how agriculture can be more productive and sustainable in the rain forest (Denevan et al. 1984; Taylor 1988; Hecht 1989; Dufour 1990). Deforestation destroys the home and the resources of these peoples, and exposes them to the diseases and other ills of the modern world (Taylor 1988).

Tropical rain forests also provide "ecosystem services." When rain falls on a forest, most of the water returns to the atmosphere through evaporation and transpiration from the vegetation; less reaches the groundwater and streams, and this arrives slowly because of the retention of moisture by the soil and litter. If forest is replaced by pasture or cropland, much less water returns to the atmosphere, less is absorbed because of soil compaction, and much more water rapidly enters streams. On a local scale this results in soil erosion, flooding, silting of rivers and reservoirs, changes in water quality, and drying up of streams during dry periods. When an entire region is deforested, the reduced return of water to the atmosphere causes a decline in regional rainfall. Furthermore, large-scale deforestation, by greatly reducing the heat-absorbing processes of evaporation and transpiration, may affect global circulation patterns and climate (Molion 1989).

Finally, there is the "greenhouse connection" of TRFs. Carbon dioxide is the most important of the greenhouse gases, those that trap heat leaving the earth's surface. Although most of the elevation in CO_2 emissions into the atmosphere is attributable to fossil fuel combustion, about 33 percent is due to land-use changes, primarily tropical deforestation (World Resources Institute 1990, 109).

Nearly half of the world's TRFs have been destroyed (mostly during the past fifty years), and the area being cleared each year is equivalent to about 2 percent of what remains (Myers 1991). Although environmentalists in developed countries have called for the preservation of remaining rain forests, most students of this issue doubt that more than a small percentage of the remaining forest area will be effectively preserved. The pressure to convert TRF to agriculture or pasture, or to exploit its timber or mineral resources, is due in part to increases in population growth, per capita consumption (in both developed and developing nations), and concentration of more productive farmland in the hands of the wealthy few. Other factors behind deforestation are the financial interests of individuals, corporations, and national governments, and the necessity of cultivating export earnings to service external debt. These root causes are exacerbated by misguided government and international development policies. Exploitation provides mostly short-term gains, however. After the most valuable timber has been removed, and the remainder burned to provide nutrients for a few harvests of crops, most TRF soils are too infertile to support agriculture (Whitmore 1990).

To accommodate both the short-term economic pressure for exploitation and the long-term need to protect the biodiversity, native cultures, and the ecosystem services of TRFs, "sustainable development" has received a great deal of attention in recent years (e.g., Anderson 1990a). The concept of sustainable development describes economic transformations that optimize societal benefits available in the present without jeopardizing the potential for benefits in the future (Goodland 1990; Dahlberg 1991). Although an underlying assumption of "sustainability" is that resource use can remain productive indefinitely in a given area, I do not accept Turner's claim that "sustainability" assumes that "the essential feature of Nature is homeostasis." Sustainable resource use emphasizes human manipulation of ecosystems, not stasis.

Sustainable development options proposed for TRFs include ecotourism (Gradwohl and Greenberg 1988, 66-67), sustainable agriculture, especially agroforestry (Ewel 1986; Montagnini 1990; Peck 1990), extractive reserves for nontimber forest products (Fearnside 1989; Browder 1992), and natural forest management for timber (Mergen and Vincent 1987). Given the prospect of its widespread application, natural forest management deserves special consideration.

Natural Forest Management

Natural forest management (NFM) involves the controlled harvesting of timber along with manipulations of the remaining plants intended to increase the future commercial value of the forest (Schmidt 1987). It contrasts with both monocultural tree plantations and "high-grading" or "creaming," the usual mode of logging in TRFs. In high-grading, the largest trees of the most valuable species are removed with no regard to damage to other trees or the soil or subsequent use of the land. In several cases, high-grading has led to the extinction of the selected tree species over vast areas (Heinsdijk and De Miranda Bastos, cited in Fox 1976). The major challenge of NFM is how to maintain or enhance the representation of commercially valuable tree species where they account for only a small proportion of the trees. An illustration of how few trees in a TRF are commercially valuable comes from a national forest in Peru, where only eleven of the three hundred tree species were commercially utilized in 1974 (UNDP/FAO 1979).

The two classes of natural forest management are monocyclic felling and polycyclic felling (Buschbacher 1990; Whitmore 1990). In monocyclic felling, all salable trees are cut in a single operation and the second harvest does not take place until a new generation of seedlings has reached marketable size. Monocyclic felling is appropriate where the commercial species can regenerate well in the open environment of a heavily logged forest. In polycyclic felling, a single area is subjected to selective logging every twenty to thirty-five years. The loggers harvest only mature trees and leave medium and small trees growing. This system makes use of the intermediate-sized trees that are spared in the felling. Both systems usually include "silvicultural treatments": the cutting or poisoning of vines and nonvaluable trees that are expected to compete with the valuable timber species.

Monocyclic Felling

The classic example of a monocyclic felling system in TRF is the Malayan Uniform System, developed for the dipterocarp forests of lowland peninsular Malaysia after World War II (Buschbacher 1990; Whitmore 1990). These forests are atypical among TRF in that a high proportion of the species, including many in the family Dipterocarpaceae, have valuable wood and these can be grouped by their wood properties into a few marketable groups. Thus it is profitable to remove a large proportion of the large trees in the initial harvest. Removing the canopy favors light-demanding, fast-growing trees, which in general are of low commercial value. In peninsular Malaysia, however, seedlings of many of the valuable dipterocarps grow well under high light conditions, if they are established before logging. These seedlings can be harvested after about seventy years. Such seedling establishment occurs naturally every several years after a "mast-fruiting," when most of the dipterocarp species produce large numbers of seeds. The first harvest must be timed within a couple years after a mast, before seedlings die from lack of sunlight. Silvicultural treatments are carried out every ten years.

Although the Malayan Uniform System seemed to have been working after twenty years, the lowlands it was designed for were subsequently converted to agriculture. Forestry then shifted to hill forests, using the Malayan Uniform System, but results have been disappointing because of variability in the abundance of seedlings of commercial species (Tang 1987).

Strip clear-cutting

An alternative type of monocyclic management is being tried in the Peruvian Amazon. In the "strip clear-cutting" system, long, narrow strips (30–40 m wide) are cleared and attempts are made to use all species and sizes of trees (Hartshorn 1989). To minimize soil damage, logs are extracted by oxen rather than tractors. The clear-cuts are permitted to regenerate naturally and it is expected that a second harvest will be possible after thirty to forty years. Logging is not carried out on swamps, slopes, or streambanks, and the remainder is managed so a few widely spaced strips are harvested each year. The spacing is intended to ensure that each cleared strip is bordered by mature and/or regenerating forest, which can serve as a source of seeds for regeneration. The expectation is that fast-growing species with good quality timber ("light hardwoods") will grow well in these strips, because many of these species regenerate in natural tree-fall gaps, and the strip clear-cut is thought to simulate a tree-fall gap. I have been studying regeneration in strips of this sort since 1988.

Polycyclic Felling

A well-documented example of polycyclic NFM is the CELOS Management System developed in Suriname (Jonkers and Schmidt 1984; de Graaf 1991). About one-third of the tree species are considered "commercial" and these account for 20 to 30 percent of the wood volume in large trees. The initial harvest is selective, taking only about 10 percent of the volume of large trees (hence one-half to two-thirds of the volume of "commercial" species). Damage is reduced by inventorying the trees before harvest and preparing an exploitation plan that minimizes roads and skid trails. In order to enhance the growth of remaining trees of commercial species, large noncommercial trees are poison-girdled one or two years after logging. Over the next twenty years the area is reentered twice to poison-girdle noncommercial trees assumed to be competing with commercial trees. A second, lighter harvest occurs after twenty years. Additional cycles of poisoning and harvest are envisioned.

Sustainability

Whether NFM is "sustainable" at a particular site depends not only on whether the growth rate of valuable trees matches the rate at which they are harvested, but also on economic feasibility, profitability compared to alternative land uses, and social and political factors. Economic analyses have generally shown NFM in TRF to be less profitable than alternative land uses such as monocultural plantations, but this conclusion is largely a function of the discount rate, an economic construct that devalues future costs and benefits, because of the delay in "payback" from silvicultural treatments (Leslie 1987). More important, such "economic" analyses are really only "financial" analyses; a true economic appraisal would have to consider non-revenue-producing benefits and external effects, such as the watershed protection and other ecosystem services provided by forests (Leslie 1987).

Social factors also impinge on whether a forestry project is sustainable. Perhaps the most important of these is land tenure (Buschbacher 1990). If individuals, communities, or institutions have no long-term stake in a forest, they are unlikely to forgo short-term profits for long-term sustainability. Furthermore, once roads have been built to extract the first timber harvest, a forest is susceptible to invasion by agriculturalists. The Nigerian Tropical Shelterwood System, a polycyclic system developed in 1950, was largely abandoned in the 1970s when Nigeria decided to replace managed natural forests with timber plantations, oil palm and cocoa plantations, and field crops in order to feed its growing population and increase export revenue (Lowe 1978). Only 2 percent of Nigeria remained reserved for forest management in 1977.

Although the ecological, economic, and social obstacles to sustainability are recognized, NFM is considered by many to be crucial to conservation plans in nations with TRF, because managed forests have greater conservation value than agricultural land (FAO 1985; Leslie 1987;

Schmidt 1987; Gradwohl and Greenberg 1988; Anderson 1990b; Busch-bacher 1990; de Graaf 1991). Let us assume that large tracts of TRF are dedicated to NFM, and that pressures to convert these managed forests to agriculture are resisted. What will be the nature of these managed forests?

The Nature of the Managed Forest

We cannot "judge the ecological viability" of a landscape by our "innate aesthetic abilities," as Turner asserts in this volume. To many people, a well-landscaped golf course or a weeded tree plantation is more aesthet-ically pleasing than the tangle of a rain forest, yet each golf course or tree plantation harbors only a tiny fraction of the number of species of a TRF, and they are maintained only with inputs of pesticides, fertilizers, and machine or manual labor. If we want to judge ecological viability, we must study ecology. The restorationist program, which William Jordan describes in this volume in his essay " 'Sunflower Forest': Ecological Restoration as the Basis for a New Environmental Paradigm," rightly emphasizes the role of preserved ecosystems, rather than aesthetic pref-erences, as models for anthropogenic ecosystems.

Before discussing the specific impacts of NFM, I stress that most of the species found in any area of TRF are *rare*. For example, in a 23-ha (0.09 mi^2) block of Malayan TRF, Poore (cited in Ng 1983) reported that the 2,663 mature trees belonged to 377 species. Of these tree species, 307 (81 percent) were represented by only one to ten trees each! Small popu-lations of a species are at risk of extinction, and any reduction in abun-dance increases the probability of extinction (Shaffer 1981). Those pop-ulations that do not decline all the way to extinction will suffer an inevitable loss of genetic diversity, which will reduce their ability to adapt to environmental change and natural enemies (and hence their abil-ity to persist in the long term) and their utility to modern agriculture and medicine.

In discussing how NFM will change a forest, we must acknowledge that a stated goal of NFM is to increase the proportion of commercially valuable tree species. Some commercial species are likely to decline be-cause (1) they are not well represented as seedlings or larger individuals at the time of the initial logging; (2) they are not able to germinate and es-tablish in the postlogging environment; and (3) they are unlikely to sprout from cut stumps. With respect to the third factor, our studies in Peru are revealing that trees in some plant families (e.g., Lecythidaceae and Vochysiaceae) usually sprout, and those in other families (e.g., Myristicaceae and Melastomataceae) rarely do (Gorchov et al. 1993).

To the extent that a management plan is successful, and commercial tree species, as a group, increase, then noncommercial tree species will decrease. In most NFM systems, such as the Malayan Uniform System and the CELOS Management System, noncommercial trees that are inferred to be competing with commercial trees are killed by poison-girdling or other means. Although the strip clear-cut system initially did not incorporate such "liberation thinnings," the preponderance of noncommercial pioneer species in the regenerating strips (Hartshorn 1988; Gorchov et al. 1993) has necessitated plans for liberation thinnings (W. Pariona, personal communication). Some noncommercial species are likely to thrive under the environmental conditions of the managed forest, but those likely to decline to local extinction include those with the same characteristics that predispose commercial species to decline: rarity, low seedling establishment in cleared areas, and low resprouting potential.

Harvest and thinning are not the only sources of mortality. When each large tree is felled, numerous other trees are broken or damaged. For example, one study in Borneo estimated that for each large tree felled, seventeen other trees were killed, and 41 percent of the survivors received crown or branch damage (Abdulhadi et al. 1981, cited in Ng 1983). A study of selective logging in the Brazilian Amazon found that 3 percent of the trees were felled but another 47 percent were uprooted by bulldozers or crushed by the fall of harvested trees (Uhl et al. 1989). Extraction of logs causes further damage to vegetation and soils, especially when carried out by mechanized vehicles (Wyatt–Smith 1987). Wounds often expose trees to infection by insects and fungi. The strip clear-cut system is less damaging than other systems, as small trees are felled and harvested before larger trees, directional felling avoids damage to plants outside of the strip, and teams of oxen drag out the logs along paths narrower than those for vehicles. Nonetheless, those species of trees, shrubs, and forest-floor herbs that are not able to withstand the damage, or sprout from broken stems, are likely to decline.

Over a time scale longer than the rotation time of twenty to eighty years, additional factors will influence which tree species prosper and which decline to extinction. Species with short generation times (able to reach reproductive age and produce seeds quickly) will be favored over those unable to reproduce in between silvicultural treatments. Trees that require cross-pollination to set seeds will fail to reproduce if they are too far from other adults of the same species, as is more likely after logging and thinnings. This problem is compounded for species that are dioecious (individual trees are either male or female): if pollinators do not transport pollen from one to the other, no seeds are produced.

Foresters focus on trees, but other plants will also become more common or rare in response to NFM. The tendency for woody vines to increase after logging is well known and often great enough to hinder tree growth (Putz 1985). I predict that epiphytes (plants that grow on other plants and are not rooted in the ground) will in many cases suffer from NFM. The number of epiphytes can be staggering; over 350 species have been recorded in only 1,500 hectares in Costa Rica, 25 percent of all plant species (Hammel 1990). Epiphytic species likely to decline in a managed forest are those that are specialized on the largest trees; in polycyclic systems large trees are generally selected for felling and in monocyclic systems few if any trees will be allowed to grow long enough to reach these sizes. Epiphytic species with long generation times, or those that depend on the high humidity beneath a dense canopy, are also likely to go extinct. It is unlikely that vulnerable species such as these could adapt to environmental changes fast enough to avoid extinction. Turner has no basis for his assertion that "higher" species evolve more rapidly; the species with the shortest generation times (for example, microbes and insects) evolve the fastest.

The effects of forest management will not be limited to plants. Some animal species will be negatively impacted by forest management via changes in nesting sites, food availability, predation, disease, and microclimate (Johns 1985). Removal of defective (i.e., broken, damaged, leaning) trees, common in forest management, may take away the only nesting places for certain species of small mammals, hornbills, and parrots (Johns 1985; Wyatt-Smith 1987). Animals with specialized diets are at greater risk of extinction than more generalized species. Many animal species are dependent on particular plant species, if not over the animal's entire life cycle than at least during a particular time of the year. If that plant species goes extinct, the animal species that depend on it will follow. Some plant species are "keystone resources" for many species of animals (Terborgh 1986). After twelve years of research at a field station in Amazonian Peru, Terborgh and his coworkers have documented that a dozen tree species, less than 1 percent of the plant species at the site, provide most of the fruits and nectar that sustain the primates, many other nonflying mammals, and many bird species during the annual dry season. Similarly, Howe (1984) argues that *Casearia corymbosa* and *Virola sebifera* are "pivotal species" for fruit-eating animals in TRFs in Costa Rica and Panama, respectively. If one or more of the plant species providing such a resource should go extinct in a managed forest, numerous bird and mammal species would follow.

Extinction of animal species is not a self-contained problem. Many of the animal species that feed on nectar, pollen, fruits, or seeds serve as dispersal agents of pollen or seeds. Many tropical plants are pollinated by only a single species of animal, so if that animal goes extinct there will be no cross-pollination, no seed set, no reproduction, and hence eventual extinction.

The vast majority of tropical trees depend on animals to disperse their seeds (Howe and Smallwood 1982). Although few if any plant species are specialized on a single animal species for seed dispersal (Wheelwright and Orians 1982), very few have been studied adequately to quantify their dependence on particular dispersal agents. The tree whose dispersal has been best studied, *Virola surinamensis*, relies on only three bird species to disperse seeds more than 20 m, where their probability of escaping seed predation by weevils is greatly enhanced (Howe and Schupp 1985). Because seeds and seedlings close to adult trees of the same species are, in most cases, more likely to be eaten by insects or mammals, or killed by fungal diseases (Clark and Clark 1984), the animals that disperse its seeds may be essential to the persistence of a plant's population. Seeds of some tropical tree species will not even germinate unless they have been separated from the surrounding pulp or aril, as occurs in the digestive system of seed-dispersing animals (Ng 1983).

Seed dispersal takes on added importance in a managed forest, where the established trees of various species may be depleted within logged areas because of harvest, damage during felling or extraction, or poisoning. Local reestablishment would depend on new seedlings' germinating either from dormant seeds in the soil or seeds dispersed in from adjacent less-damaged forests. In TRF seed dormancy is largely confined to pioneer trees, vines, and weedy herbs (Janzen and Vasquez-Yanes 1991); the "seed bank" cannot replenish the population of most late successional and climax forest tree species.

Tree species dispersed by wind and bats will be better able to reseed logged areas than those dispersed by birds or arboreal mammals such as primates. Although birds are capable of flying across disturbed areas and excreting seeds in flight or while perched, they rarely do. In the 30-m-

Table 1. Number of seeds per m^2 of two bird-dispersed plant species recovered in seed traps at three different distances from the edge of a 30-m-wide strip and in intact forest 25 m from the strip over the first year after strip-cutting in the Peruvian Amazon. For explanation of location and number of traps, see figure 1.

Species	Distance from forest edge			
	12.5 m	7.5 m	2.5 m	Intact forest
Miconia minutiflora	11.6	34.9	388.1	582.4
Ossaea cucullata	0.8	5.9	12.1	189.9

wide strip we have studied most intensively in the Peruvian Amazon, seed traps placed 12.5 m from the forest averaged only two-fifths to one-tenth as many bird-dispersed species as traps in the forest during each of the first two years after clear-cutting (see figs. 1 and 2). Over the first year after clearing, traps near the center of the strip contained only 2 percent as many seeds of the most numerous bird-dispersed species as traps in the forest, and only 0.4 percent as many seeds of the second most numerous (table 1).

Those bird-dispersed seeds that did reach the interior of the strip in our study were tiny seeds of noncommercial shrubs and pioneer (colonizing) trees (Gorchov et al. 1993). While many of the commercially valuable large trees in the surrounding forest have seeds that are dispersed by birds, these are adapted for dispersal by larger birds such as toucans and cotingids. Despite numerous hours of observation, we have not observed those birds crossing over the strips, though they are present in the surrounding forest. The reluctance of such dispersal agents to venture into cleared areas may be due to the low stature of the vegetation, but regardless of the cause it questions the assumption that seeds will somehow disperse from the forest into cleared strips.

Ecosystem-level changes will also occur as a TRF is converted to a managed forest. One certain change is a reduction in the pool of mineral nutrients, as biomass is removed from the site in the form of logs or sawn wood. Such nutrient loss can be minimized by debarking logs before extracting them, as bark has higher levels of nutrients than wood does. Nevertheless, some nutrients will be removed with the logs, and because

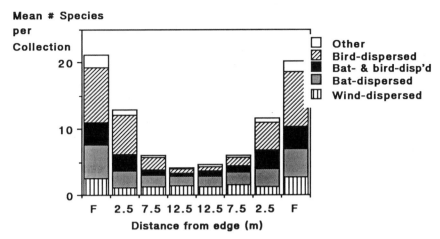

Figure 1. Mean number of plant species represented in seed traps per semimonthly collection in a logged strip in the Peruvian Amazon over the first year after clearing (June 1989-June 1990). Each bar represents a set of sixteen traps (total area 4 m²) located the same distance from the forest/strip edge: "F" traps were in the forest 25 m from the edge; "2.5" traps were in the strip 2.5 m from the forest; "7.5" and "12.5" traps were in the strip 7.5 and 12.5 m from the forest, respectively. Left to right on the graph corresponds to west to east across the strip, which measured 30 m east-west and 150 m north-south. Plant species are categorized by dispersal mode. Reprinted with permission from *Vegetatio* (Gorchov et al. 1993).

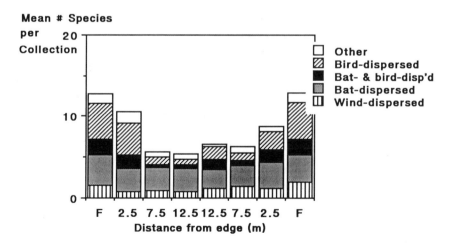

Figure 2. As in figure 1, except data are means for the second year after clearing (June 1990-June 1991).

tropical soils generally have very low levels of nutrients, these losses are likely to have some effect (Poels 1987, cited in Buschbacher 1990).

These losses of biodiversity have been acknowledged by foresters. In their discussion of the CELOS Management System, Jonkers and Schmidt (1984) recognize that "the CELOS silvicultural system interferes in the ecosystem in many ways. . . . non-commercial species which do not complete their life cycle below the diameter limit of 20 cm are likely to be eliminated completely" (295). For the strip clear-cut system, Hartshorn (1989) admits, "We assume that those species not occurring in the excluded (unlogged) habitats or that cannot become sexually mature within the rotation period will be lost from the managed forests" (568). For the Malayan Uniform System and its variants, Wyatt-Smith (1987) predicts that "the managed lowland forests in the future will not resemble directly the primary forests in either structure or species composition and diversity" (20). Johns (1985), however, predicts that large areas of managed forest will support (if hunting is restricted) many animal species, including top predators, that would go extinct in small patches of primary forest in a matrix of nonforest land.

This is not to suggest that human modification of forests is only a modern phenomenon. Forests in the estuary of the Amazon River have been modified by centuries of "tolerant" management by local people, favoring species that provide timber, edible fruits, fibers, latex, and medicinals, and thinning less desirable species (Anderson 1990c). The prevalence of economically important tree species in tropical forests in southern Mexico is attributed to their origin as abandoned pre-Columbian "forest gardens," where valuable species were planted or encouraged, other species were eliminated, and the forest protected from fire and destruction (Gomez-Pompa and Kaus 1990). In southern Nigeria, forests growing on land with evidence of agriculture two hundred years in the past have more tree species than forests believed to be virgin (Jones 1956). These studies show that modified forests may be structurally similar to forests with little human intervention, and may retain a high diversity of plant species.

This conclusion does not mean that there is no such thing as a pristine forest, or that we cannot distinguish between "natural" and "anthropogenic" ecosystems. Forests in areas of historically very low population density, such as the Amazonian uplands, are believed to show little or no influence of human intervention. In fact, it is the lower species diversity and higher frequency of a few valuable species in the "traditionally managed" forests that suggest or support the hypothesis that they are anthropogenic. Forests managed by modern NFM will develop into even more extreme "simplifications" of natural forests.

The Value of Managed Forests

Will these managed, semiwild forests be equivalent or inferior to natural forests? From a financial perspective, and probably from a true economic perspective, they will be superior, as long as the principal objective of increased representation of valuable species is met. However, we cannot make the assumption made by so many policymakers in recent years, that sustainable development will automatically protect biodiversity (Redford and Sanderson 1992). Managed forests are likely to be equivalent, or slightly inferior, in terms of the ecosystem services they provide. In contrast to conversion of the land to agriculture or pasture, NFM will retain forest cover over the land and therefore minimize soil erosion and disruption of the regional hydrological cycle. Over a larger scale, maintenance of the hydrological cycle should avoid major changes to atmospheric circulation patterns. To the extent soil and roots are not damaged during harvest, nutrients will be recycled within the forest, avoiding the impoverishment of soil associated with agriculture and pasture.

From the point of view of atmospheric CO_2 and global warming, NFM is attractive because the managed forests will have greater biomass (and hence more carbon locked up in organic compounds) than pastures, agriculture fields, or agroforestry plots. Nevertheless, this biomass will be lower than in natural forests, even at the end of a growing cycle (before a harvest), because of the lower canopy height of the managed forest. Furthermore, at any one time only a portion of the managed forest will be at this stage.

Although managed forests are likely to perform their ecosystem services nearly as well as unlogged forests, they will not support the same biodiversity. For the reasons outlined earlier in this essay, we can expect plant and animal species to go extinct, and these extinctions will cause further extinctions. We cannot predict how many species will go extinct because of the complexity of the ecological interactions and our ignorance of them. Although extinction is part of the history of life on earth, the rate of extinction predicted from tropical deforestation greatly exceeds "normal" extinction rates. Even the cataclysmic mass extinction that marks the end of the Cretaceous is believed by several paleontologists to have taken place over many thousands of years (Futuyma 1986, 364), rather than "swifter than any human effect on the environment," as Turner claims.

The Limits to Management of Biodiversity

The conservation of a species of particular importance might be achieved, if its ecology were well known, by modifications of the management plan, just as we have learned to manage tall-grass prairies (Jordan, " 'Sunflower Forest' "). But it is foolish to presume that humans could manage a TRF to prevent extinctions. First, we understand the ecology of only a tiny minority of TRF species, and, with so much research needed on the others, we can hope to add to this list only incrementally. Second, we know even less about how individual species will respond to the management tools we could try. Third, the management techniques that may be necessary to conserve one particular species are likely to have negative consequence on other species. Even with complete knowledge, we could not expect to manage for more than a handful of species at a time. Finally, the cost of such comprehensive research, management, and monitoring would be prohibitive. Jordan's claim that in the long run the best natural areas will be restored sites, rather than preserved sites, though plausible for relatively simple ecosystems such as prairies, is outlandish for systems as complex as rain forests.

There is one way to conserve most of the species of the TRF: *preservation* of large tracts with minimal interference from humans. In this volume Turner and Jordan rightly point out that a preserved ecosystem will not remain in its present state indefinitely, but rather will change in response to various forces including, at the slowest time scale, evolution. This should not cause us to ridicule preservation as an impossibility, however, but to recognize that by preservation we mean "protect from degradation," not "preserve in precisely the present state."

Preserved areas of TRF could probably sustain some income-generating human activity, such as ecotourism, controlled extraction of nontimber forest products, and limited hunting and fishing. The case for a nation's leaving large tracts in such relatively "unproductive" land use will be strengthened if that nation's demand for timber and for the revenue generated by export of timber is met by sustained production in other areas, either by NFM or tree plantations. Such sustainable development should be designed to benefit the people living in the affected area. Because a managed forest provides superior ecosystem services and maintains higher biodiversity than agriculture, natural forest management is a desirable component of national and international conservation strategy. If the goal is to avoid massive extinctions, however, we cannot fail to acknowledge that *preservation* is the essential core strategy.

REFERENCES

Anderson, A. B., ed. 1990a. *Alternatives to Deforestation: Steps toward Sustainable Use of the Amazon Rain Forest*. New York: Columbia University Press.

————. 1990b. "Deforestation in Amazonia: Dynamics, Causes, and Alternatives." In A. B. Anderson, ed., *Alternatives to Deforestation: Steps toward Sustainable Use of the Amazon Rain Forest*. New York: Columbia University Press. 3-23.

————. 1990c. "Extraction and Forest Management by Rural Inhabitants in the Amazon Estuary." In A. B. Anderson, ed., *Alternatives to Deforestation: Steps toward Sustainable Use of the Amazon Rain Forest*. New York: Columbia University Press. 65-85.

Browder, J. O. 1992. "The Limits of Extractivism." *BioScience* 42: 174-82.

Buschbacher, R. J. 1990. "Natural Forest Management in the Humid Tropics: Ecological, Social, and Economic Considerations." *Ambio* 19: 253-58.

Clark, D. A., and D. B. Clark. 1984. "Spacing Dynamics of a Tropical Rain Forest Tree: Evaluation of the Janzen-Connell Model." *Amer. Nat.* 124: 769-88.

Dahlberg, K. A. 1991. "Sustainable Agriculture—Fad or Harbinger?" *BioScience* 41: 337-40.

de Graaf, N. R. 1991. "Managing Natural Regeneration for Sustained Timber Production in Suriname: The Celos Silvicultural and Harvesting System." In A. Gomez-Pompa, T. C. Whitmore, and M. Hadley, eds., *Rain Forest Regeneration and Management*. Man and the Biosphere Series, vol. 6. Paris: UNESCO, and Park Ridge, N.J.: Parthenon Publishing Group. 393-405.

Denevan, W. M., J. M. Treacy, J. B. Alcorn, C. Padoch, J. Denslow, and S. F. Paitan. 1984. "Indigenous Agroforestry in the Peruvian Amazon: Bora Indian Management of Swidden Fallows." *Interciencia* 9: 346-57.

Dufour, D. L. 1990. "Use of Tropical Rainforests by Native Amazonians." *BioScience* 40: 652-59.

Ewel, J. J. 1986. "Designing Agricultural Ecosystems for the Humid Tropics." *Annu. Rev. Ecol. Syst.* 17: 245-71.

FAO. 1985. *Tropical Forestry Action Plan, Committee for Forest Development in the Tropics*. Rome: Food and Agriculture Organization of the United Nations.

Fearnside, P. M. 1989. "Extractive Reserves in Brazilian Amazonia." *BioScience* 39: 387-93.

Fox, J. E. D. 1976. "Constraints on the Natural Regeneration of Tropical Moist Forest." *Forest Ecol. Mgt.* 1: 37-65.

Futuyma, D. J. 1986. *Evolutionary Biology*. 2d ed. Sunderland, Mass.: Sinauer.

Gomez-Pompa, A., and A. Kaus. 1990. "Traditional Management of Tropical Forests in Mexico." In A. B. Anderson, ed., *Alternatives to Deforestation: Steps toward Sustainable Use of the Amazon Rain Forest*. New York: Columbia University Press. 45-64.

Goodland, R. 1990. "Environmental Sustainability in Economic Development—with Emphasis on Amazonia." In R. Goodland, ed., *Race to Save the Tropics*. Washington, D.C.: Island Press. 171-89.

Gorchov, D. L., F. Cornejo, C. Ascorra, and M. Jaramillo. 1993. "The Role of Seed Dispersal in the Natural Regeneration of Rain Forest after Strip–cutting in the Peruvian Amazon." *Vegetatio* 107–8: 339–49.

Gradwohl, J., and R. Greenberg. 1988. *Saving the Tropical Forests*. London: Earthscan Publishing Ltd.

Hammel, B. 1990. "The Distribution of Diversity among Families, Genera, and Habit Types in the La Selva Flora." In A. H. Gentry, ed., *Four Neotropical Forests*. New Haven: Yale University Press. 75–84.

Hartshorn, G. S. 1988. "Natural Regeneration of Trees on the Palcazu Demonstration Strips." Washington, D.C.: Forestry Support Program, U.S. Department of Agriculture.

———. 1989. "Application of Gap Theory to Tropical Forest Management: Natural Regeneration on Strip Clear–cuts in the Peruvian Amazon." *Ecology* 70: 567–69.

Hecht, S. B. 1989. "Indigenous Soil Management in the Amazon Basin: Some Implications for Development." In J. O. Browder, ed., *Fragile Lands of Latin America: Strategies for Sustainable Development*. Boulder, Colo.: Westview Press. 161–81.

Howe, H. F. 1984. "Implications of Seed Dispersal by Animals for Tropical Reserve Management." *Biol. Cons.* 30: 261–81.

Howe, H. F., and J. Smallwood. 1982. "Ecology of Seed Dispersal." *Annu. Rev. Ecol. Syst.* 13: 201–28.

Howe, H. F., and E. W. Schupp. 1985. "Early Consequences of Seed Dispersal for a Neotropical Tree (*Virola surinamensis*)." *Ecology* 66: 781–91.

Janzen, D. H., and C. Vasquez-Yanes. 1991. "Aspects of Tropical Seed Ecology of Relevance to Management of Tropical Forested Wildlands." In A. Gomez-Pompa, T. C. Whitmore, and M. Hadley, eds., *Rain Forest Regeneration and Management*. Man and the Biosphere Series, vol. 6. Paris: UNESCO, and Park Ridge, N.J.: Parthenon Publishing Group. 137–57.

Johns, A. D. 1985. "Selective Logging and Wildlife Conservation in Tropical Rain-Forest: Problems and Recommendations." *Biol. Cons.* 31: 355–75.

Jones, E. W. 1956. "Ecological Studies on the Rain Forest of Southern Nigeria. IV. The Plateau Forest of the Okomo Forest Reserve. Part II. The Reproduction and History of the Forest." *J. Ecology* 44: 83–117.

Jonkers, W. B., and P. Schmidt. 1984. "Ecology and Timber Production in Tropical Rainforest in Suriname." *Interciencia* 9: 290–97.

Leslie, A. J. 1987. "A Second Look at the Economics of Natural Forest Management Systems in Tropical Mixed Forests." *Unasylva* 39: 46–58.

Lowe, R. G. 1978. "Experience with the Tropical Shelterwood System of Regeneration in Natural Forest in Nigeria." *Forest Ecol. Mgt.* 1: 193–212.

Mergen, F., and J. R. Vincent, eds. 1987. *Natural Management of Tropical Moist Forests: Silvicultural and Management Prospects of Sustained Utilization*. New Haven: School of Forestry and Environmental Studies, Yale University.

Miller, K., and L. Tangley. 1991. *Trees of Life*. Boston: Beacon Press.

Molion, L. C. B. 1989. "The Amazonian Forests and Climatic Stability." *The Ecologist* 19: 211–13.

Montagnini, F. 1990. "Ecology Applied to Agroforestry in the Humid Tropics." In R. Goodland, ed., *Race to Save the Tropics*. Washington, D.C.: Island Press. 49–58.

Myers, N. 1983. *A Wealth of Wild Species: Storehouse for Human Welfare*. Boulder, Colo.: Westview Press.

———. 1991. "Tropical Deforestation: The Latest Situation." *BioScience* 41: 282.

Ng, F. S. P. 1983. "Ecological Principles of Tropical Lowland Rain Forest Conservation." In S. L. Sutton and T. C. Whitmore, eds., *Tropical Rain Forest: Ecology and Management*.

Special Publication No. 2 of the British Ecological Society. Oxford and Boston: Blackwell Scientific Publishers. 359-75.

Peck, R. B. 1990. "Promoting Agroforestry Practices among Small Producers: The Case of the Coca Agroforestry Project in Amazonian Ecuador." In A. B. Anderson, ed., *Alternatives to Deforestation: Steps toward Sustainable Use of the Amazon Rain Forest*. New York: Columbia University Press. 167-80.

Plotkin, M. J. 1988. "The Outlook for New Agricultural and Industrial Products from the Tropics." In E. O. Wilson, ed., *Biodiversity*. Washington, D.C.: National Academy Press. 106-16.

Posey, D. A. 1990. "Intellectual Property Rights: What Is the Position of Ethnobiology?" *J. Ethnobiol.* 10: 93-98.

Putz, F. E. 1985. "Woody Vines and Forest Management in Malaysia." *Commonw. For. Rev.* 64: 359-65.

Redford, K. H., and S. E. Sanderson. 1992. "The Brief, Barren Marriage of Biodiversity and Stability." *Bull. Ecol. Soc. Amer.* 73: 36-39.

Schmidt, R. 1987. "Tropical Rain Forest Management: A Status Report." *Unasylva* 39 (156): 2-17.

Shaffer, M. L. 1981. "Minimum Population Sizes for Species Conservation." *BioScience* 31: 131-34.

Soejarto, D. D., and N. R. Farnsworth. 1989. "Tropical Rain Forests: Potential Source of New Drugs?" *Perspectives in Biol. & Medicine* 32: 244-56.

Tang, H. T. 1987. "Problems and Strategies for Regenerating Dipterocarp Forests in Malaysia." In F. Mergen and J. R. Vincent, eds., *Natural Management of Tropical Moist Forests: Silvicultural and Management Prospects of Sustained Utilization*. New Haven: School of Forestry and Environmental Studies, Yale University. 23-41.

Taylor, K. I. 1988. "Deforestation and Indians in Brazilian Amazonia." In E. O. Wilson, ed., *Biodiversity*. Washington, D.C.: National Academy Press. 138-44.

Terborgh, J. 1986. "Community Aspects of Frugivory in Tropical Forests." In A. Estrada and T. H. Fleming, eds., *Frugivores and Seed Dispersal*. Dordrecht: Dr W. Junk Publishers. 371-84.

Uhl, C., D. Nepstad, R. Buschbacher, K. Clark, B. Kauffman, and S. Subler. 1989. "Disturbance and Regeneration in Amazonia: Lessons for Sustainable Land-use." *The Ecologist* 19: 235-40.

UNDP/FAO. 1979. *Demonstración de manejo y utilización integral de bosques tropicales. Peru. Plan de manejo para el Bosque Nacional Alexander von Humboldt*. Technical Report 1. FO:DP/PER/71/551. Rome: Food and Agriculture Organization of the United Nations.

Vaughan, D. A., and L. A. Sitch. 1991. "Gene Flow from the Jungle to Farmers." *BioScience* 41: 22-28.

Wheelwright, N. T., and G. H. Orians. 1982. "Seed Dispersal by Animals: Contrasts with Pollen Dispersal, Problems of Terminology, and Constraints on Coevolution." *American Naturalist* 119: 402-13.

Whitmore, T. C. 1990. *An Introduction to Tropical Rain Forests*. Oxford: Clarendon Press.

World Resources Institute. 1990. *World Resources 1990-91*. New York and Oxford: Oxford University Press.

Wyatt-Smith, J. 1987. "Problems and Prospects for Natural Management of Tropical Moist Forests." In F. Mergen, and J. R. Vincent, eds., *Natural Management of Tropical Moist Forests: Silvicultural and Management Prospects of Sustained Utilization*. New Haven: School of Forestry and Environmental Studies, Yale University. 5-22.

Identifying a Strategy for Forest Restoration in the Tana River National Primate Reserve, Kenya

Kimberly E. Medley

Ecology and environmentalism are different. Environmentalists maintain a set of ideals as to how humans should sustain their link with nature, and ecologists have through successive studies sought to understand the mechanisms by which nature operates. Creativity in ecology is achieved through the questions that are asked, but the discipline remains conservative in its approach toward interpreting research results. By critically synthesizing the results from nearly one hundred years of ecological research, it is understandable why skepticism arises concerning the ability of humans to restore or invent ecosystems (e.g., McIntosh 1985). Despite the arguments of Frederick Turner and William Jordan that are presented in this book, ecological restoration is not easily applied. Natural systems, with or without humans, are a complex array of species interacting dynamically with their physical environment. Only by interpreting results with the idealistic objective of improving relationships between humans and their environment (i.e., environmentalism), or with a more narrowly defined goal, can humans ever expect to succeed in ecological restoration.

Much of the scientific debate regarding restoration may be traced to an unattainable statement initiative: the re-creation and maintenance of the spatiotemporal dynamics and functioning of an ecosystem. Restoration management serves as a less-suitable alternative to establishing "very large protected islands" and an a priori decision to exclude humans. Research has documented well the relationships among human activities, landscape fragmentation, and species extinction (Shafer 1990; Lovejoy et al. 1986). The dilemma, however, is that species are typically lost from protected islands, be-

154

cause of natural or human disturbances acting on these "small" areas (Pickett and Thompson 1978; Newmark 1987). Moreover, creating larger and larger islands is often in direct conflict with the land-resource needs of local residents (e.g., Lusigi 1981) or ecologically impossible.

If whole-ecosystem restoration is therefore not an obtainable goal, is it possible to restore sites that will protect key species in a degrading system? These species may play important roles in community functions, or be "flagship" species recognized by their aesthetic importance to humans. Restoration-project objectives are directed specifically at providing suitable *habitat* for the target species, regardless of *ecosystem* complexity. Species population sizes are the primary measure of project success. Although limited in scale, from ecosystem to habitat, site restoration may be more easily addressed and could potentially contribute toward maintaining ecological integrity at a higher level.

This essay describes a forest-restoration project for the Tana River National Primate Reserve, Kenya (TRNPR; see Medley 1990; Kinnaird et al. 1990). Restoration is proposed as a management strategy to preserve two endangered primates found only in the riverine forest patches of the lower Tana River. I outline a justification for such limited restoration, and illustrate how results from an ecological study of the forest flora can be used to identify appropriate sites and plant species. Incorporating both the scientific and humanistic perspectives well developed in this book, this case study differentiates ecological restoration at the habitat, ecosystem, and landscape levels, and summarizes how restoration may contribute toward the forest-conservation practices of this region.

The Landscape Setting

The TRNPR is located in Eastern Kenya along the floodplain section of the Tana River (fig. 1). The Tana flows from the humid highlands near Mt. Kenya and the Aberdare mountain range, through an arid-semiarid floodplain between Garissa and Garsen, and to the Indian Ocean north of Malindi. Long-term climatic records for Hola, approximately 40 km upstream from the reserve, indicate an average annual precipitation of 470 mm, no months with rainfall greater than 100 mm, and a growing season of 40 days coincident with the rains between October and December (FAO 1984). The site is equatorial with minimum and maximum temperature averages equal to 21.4° C and 33° C, respectively (Muchena 1987). The life zone is thorn woodland (*sensu* Holdridge 1967).

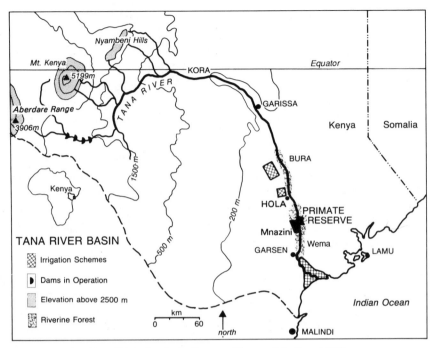

Figure 1. The Tana River basin in Eastern Kenya (Medley 1992).

Forest vegetation along the Tana River is groundwater-dependent, and its lateral extent is determined sharply by the increase in the depth to water with distance from the river (Hughes 1988). Periodic disturbances by river meanders, flooding, and human activity have created a patchy distribution of forests in a floodplain corridor that is about 1 km wide (fig. 2). Thus, the riverine forest is composed of isolated patches, which together comprise an isolated island of forest vegetation in a semiarid environment (Hamilton 1974).

Restricted almost entirely to the forest patches, the Tana River red colobus (*Colobus badius rufomitratus*) and crested mangabey (*Cercocebus galeritus galeritus*) monkeys range from the Wema/Hewani region to the northern boundary of the TRNPR (Decker and Kinnaird 1992; fig. 1). The TRNPR was established in 1976 to protect these two primates and their riverine forest habitat (Marsh 1976). It includes approximately 9.5 km² forested area in twenty-six forest patches and 60 percent of the primate populations. About 600-700 Pokomo agriculturalists live in the reserve. Although the riverine forest ecosystem is clearly dependent on the

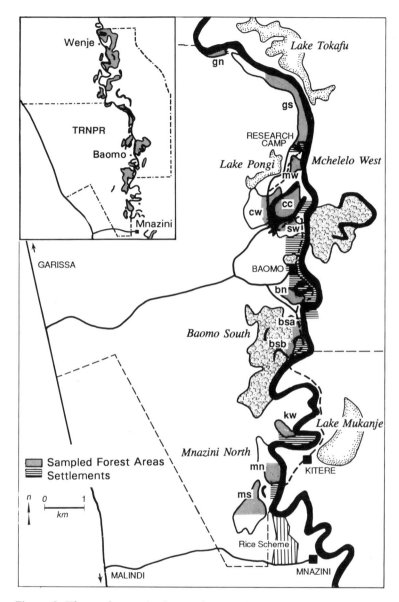

Figure 2. The study area in the south-central sector of the Tana River
National Primate Reserve (Medley 1992). The twelve forest areas
sampled include: Guru North (gn), Guru South (gs), Mchelelo West
(mw), Congolani Central (cc), Congolani West (cw), Sifa West (sw),
Baomo North (bn), Baomo South a (bsa), Baomo South b (bsb),
Kitere West (kw), Mnazini North (mn), and Mnazini South (ms). The
top inset map shows the reserve boundary and riverine forest patches
(shaded).

Tana River as a water resource, the protected area extends only a short distance (~36 km) along its present channel. The TRNPR is one of the smallest reserves in East Africa.

Environmental gradients and human activities greatly influence the distribution and ecology of the Tana River forests. Upstream from the reserve, there is a loss of important primate food resources and a general decline in forest structure. These changes may be explained by a corresponding decrease in precipitation (~300 mm at Bura) and an increase in distance from the species–rich coastal flora (Medley 1992; fig. 1). Natural forest decline, coupled with development of the Bura Irrigation Scheme and Hola Irrigation Scheme (ca. 1978), suggests that floodplain forests upstream from the TRNPR will remain unsuitable for the two primates (Hughes 1987; Ledec 1987). Downstream from the reserve, riverine forest becomes more rich in coastal plant species. These forests are best developed in the Wema/Hewani region, which is adjacent to the recent Tana River Delta Rice Irrigation Scheme. It may be possible to maintain some forest patches through implementation of this large development project, but management for forest expansion seems unlikely (Medley et al. 1989). The forests protected within the TRNPR will probably be the only habitat for the primates in the future.

Between 1976 and 1985, a resurvey of the primate populations showed an 80 percent decline in the red colobus and a 25 percent decline in the crested mangabey (Marsh 1986). Decker and Kinnaird (1992) suggest that the population declines are at least partially explained by changes in forest cover. Between 1960 and 1975, I measured from air photos a 56 percent loss in forest area and a fragmentation of five forests into fifteen forest patches in the south-central sector of the TRNPR (Medley 1993). Forest loss, fragmentation, and consequent habitat decline during this time period are explained by two primary disturbances: natural river dynamics and human forest clearing.

Forest Restoration: Project Approach and Rationale

The factors that reduce or degrade floodplain forest, coupled with the small size of the protected area, jeopardize the long-term preservation of the Tana River red colobus and crested mangabey. Their protection is dependent on the maintenance of high–quality habitat: large forests that provide some protection from the savanna-bushland, closed tree canopy greater than 10 m in height, and a variety of food resources with at least a few figs (*Ficus* spp.). Tree species composition varies from forests dominated by mchambia, *Pachystela msolo*, or a mixed composition with mniembembe, *Sorindeia madagascariensis*, and mkuru, *Diospyros mespiliformis* (Medley 1990; Decker 1989; Kinnaird 1990). Restoration aims to increase forest area, establish corridors between forest patches, and/or enrich the resources of existing forest patches (cf. Bradshaw 1987; Cairns 1988). Managed plantings of primate food resources would more specifically improve the recovery rate of their habitat (e.g., Franklin et al. 1988). Expected outcomes are increases in the primate carrying capacities of the forest patches and less isolation among existing populations (see Harris 1984; Gilpin 1987; Kinnaird 1990). The goals of the project are established at the species level with primate population sizes used as an indirect measure of project success.

In order to be successful, a restoration project should be designed to complement the regional ecology and especially the relative environmental tolerances of riverine forest plant species (Anderson and Ohmart 1985; Ashby 1987). I conducted a study to document the ecological attributes of sites and plant species appropriate for a forest restoration project (Medley 1990). I assumed that managed plantings of primate food resources would be most successful where: (a) they closely parallel or encourage the existing patterns of regeneration; (b) the hydrologic conditions are suitable for early tree establishment and growth; and (c) the plantings enrich existing forest resources. The criteria are necessarily restrictive, thereby limiting the project to a few sites that would prove "less difficult" to restore and greatly improve existing site conditions for the two primates. Plant species are selected based on their resource value to the primates, the ease at which they may be propagated, and their growth rate and tolerances at a forest site. Together these data contribute toward a management strategy that includes forest restoration within the TRNPR.

Data and Methods

Site Selection

Sampling points (n = 363) were systematically located in twelve study areas in order to quantify spatial variation within forest patches according to three site criteria: (1) evidence of forest regeneration; (2) suitable hydrologic conditions; and (3) low canopy-tree coverage (fig. 2; table 1). I

Table 1. Ecological criteria used in the selection of restoration sites in the Tana River National Primate Reserve, Kenya. Locations meeting the first two criteria are suitable for managed plantings, and identify sites where forest may be restored. Locations meeting all three criteria are suitable for enrichment plantings in a forest stand.

1. Evidence of Forest Regeneration	*Ficus sycomorus, Pachystela msolo,* and *Sorindeia madagascariensis* are important primate food resources and are characteristic of high-quality primate habitat. Sites where young trees (i.e., saplings) of one or more of these species occur show potential for habitat restoration.
2. Suitable Hydrologic Conditions	Surface soil moisture is critical for early tree establishment. Those sites that receive flood water (> 0.1 cm) in at least a two-year flood event, such as 1988, possess an acceptable hydrologic regime.
3. Low Canopy-Tree Coverage	Sites with low basal areas (< 15 m²/ha) of large trees > 20 cm in diameter are appropriate for enrichment plantings.

selected forests that were representative of the vegetation differences in the TRNPR and for which data were available on the endangered primate populations. At each point, data were compiled on the density of saplings (trees > 1 m in height and < 10 cm diameter at breast height (dbh) in 24 m^2 plots), maximum water height (cm) above the ground surface during the 1988 flood, and the basal area of canopy trees (m^2/ha of trees > 20 cm dbh, based on point-centered quadrat sampling) (Greig-Smith 1983). A geographic information system (GIS) was used to show spatial patterns within a forest patch, extrapolating from the respective sampling points, and to determine whether a locality met the criteria as a site for forest restoration (1 and 2), or as a site for forest enrichment (1, 2, and 3) (Burrough 1986; Eastman 1990; Young and Renwick 1992).

Species Selection

During a year of field research (1987-88), I grew several tree species in a nursery from seeds, stem cuttings, or forest transplants. Data were collected on germination rates and/or tree survival and growth. In addition, the growth rates (i.e., annual diameter increments) of trees > 10 cm dbh were determined for five species examined in the trials. Together these data provide information on the ease at which plant species may be propagated and projected growth rates for trees established at the restoration site.

Results

Site Selection

Approximately 10 percent of the points show establishment by one or more of the primate food resources and were saturated during the 1988 flood (fig. 3a). These localities meet criteria 1 and 2, suggesting conditions suitable for the restoration of primate habitat (table 1). Where they occur adjacent to a forest edge (e.g., Guru North, Baomo North, Baomo South, and Kitere West) or are close to another forest patch (e.g., Kitere West and Baomo South, Mnazini North and Mnazini South), restoration plantings could increase the size of a forest patch or connect two patches, respectively. The map clearly shows the potential for a corridor connection between Mnazini North and Mnazini South. Four locations are adjacent to areas under cultivation by the Pokomo (see fig. 2), suggesting that these areas could be restored to forest suitable for the primates. Approximately 4 percent of the sites suitable for restoration also have a low coverage of large trees (fig. 3b). Tree plantings within localities meeting all three criteria (e.g., Baomo South and Baomo North), though not increasing area, enrich the resource base available for the primates in a forest. Together, the two maps of figure 3 show localities where it may be easiest to increase the area of a forest patch, connect two patches via a planted corridor, or enrich the resources of an existing patch.

Species Selection

I identify two groups of tree species based on the nursery trials and plot records of their current distribution within the forests: (1) trees that grow in open areas, and (2) trees that grow in partial shade or under established

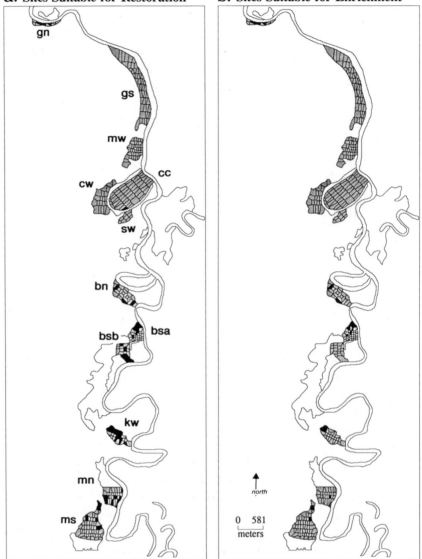

a. Sites Suitable for Restoration   **b.** Sites Suitable for Enrichment

Figure 3. Spatial analysis of the site criteria in the twelve forest areas. The black shaded areas are: (a) sites suitable for restoration, meeting criteria 1 and 2; and (b) sites suitable for forest enrichment, meeting all three criteria. Names of the forest areas are provided in figure 2.

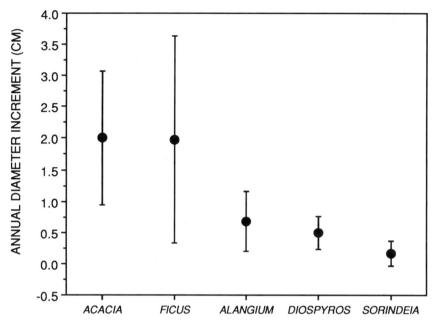

Figure 4. Mean annual growth for five tree species: *Acacia robusta* subsp. *usambarensis* (n = 22), *Ficus sycomorus* (n = 44), *Alangium salviifolium* (n = 12), *Diospyros mespiliformis* (n = 15), *Sorindeia madagascariensis* (n = 22). All individuals were > 10 cm dbh. Much of the variability in growth rates is attributable to site differences.

forest canopies. *Ficus sycomorus* and *Acacia robusta*, in the first group, grow much faster (~2 cm/year) than the more shade tolerant *Alangium salviifolium, Diospyros mespiliformis,* and *Sorindeia madagascariensis* (< 0.75 cm/year; fig. 4). Restored habitat along a forest edge or a corridor between forest patches may be quickly established by planting fast-growing pioneer trees. Later plantings of slower-growing, more-shade-tolerant trees would encourage the development of a mixed forest and potentially improve the persistence of the established community. Figs (*Ficus* spp.), transplanted from a nursery or established in situ, may be especially important to the success of the project. *Ficus sycomorus* is a primary food resource for both primates (Decker 1989; Kinnaird 1990), and results from this study show promising success in a nursery and fast growth rates in the field. The other indigenous figs appear to be of similar value to the primates and are also easily propagated from seeds. Their current rare status within the TRNPR, with mature individuals measured at only 4 out of 363 points, would be improved by successful plantings of nursery-propagated seedlings or stem cuttings.

The results from this study do not specifically address in-field trials of the selected species. On the contrary, the climatic and hydrologic conditions I described as characteristic of the region imply that early survival is precarious. Plantings should coincide with periods of soil saturation, occurring just subsequent to a flood event. Furthermore, the study has not considered all forest patches within the TRNPR, or how irrigation may improve conditions for restoration. I recommend that the project be limited to the few identified localities and implemented at a controlled, experimental level.

Implications for Conservation

In view of the critically low colobus and mangabey populations and coupled with the natural and human-related factors that reduce primate habitat within this region, forest restoration is supported as a potentially powerful management strategy. I conclude from a study of the vegetation that suitable sites are present for the establishment of primate habitat, appropriate plant species can be propagated, and habitat recovery through plantings of fast-growing species should be rapid. There is evidence that primate habitat can be successfully established or enriched at locations within the TRNPR.

Nowhere in the description of this case study have I addressed the conservation of resources at the ecosystem and landscape levels (fig. 5). The management strategy is directed at a geographic scale equivalent with the forest patch, or a locality within a forest patch. Restoration at this scale focuses on community dynamics; that is, encouraging the establishment of primate habitat. The habitat preferences of the endangered primates and their small geographic range along the river limit their value as an indicator of ecosystem biodiversity and landscape heterogeneity. The factors that influence local-regional diversity and the environmental gradients along the river are simply not addressed in the proposed strategy. Nevertheless, restoration applied at the site scale should document differences among plant-species tolerances, environmental conditions necessary for effective growth, and rates of local forest development. If applied, habitat restoration may identify functional links within the resource hierarchy that are critical for the preservation of ecosystem and landscape resources.

Furthermore, forest restoration does not set exclusion of the resident Pokomo population as an ultimatum for preservation. On the contrary, ef-

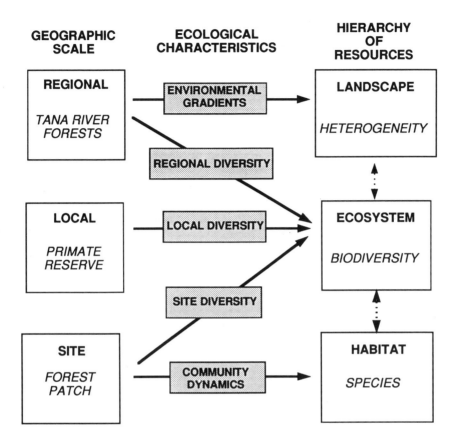

Figure 5. Schematic diagram showing the relationships among geographic scale, influencing ecological characteristics, and the consequent hierarchy of resources at the habitat, ecosystem, and landscape levels.

fective conservation within this region depends on the cooperation of the people and their contributory efforts toward reserve protection. Similar to the "Sunflower Forest" described by Jordan in the first essay of this volume, forest restoration within the TRNPR may serve as a constructive outlet for human land use that could be mutually beneficial. Wage labor provided by management or research activities compensate for limited resource use within the protected area. The ideals of environmentalism are nurtured by including residents in the conservation of a local resource. Through a combined effort, the complex ecology of an area is further understood upon implementation of a well-defined, however limited, restoration management plan.

ACKNOWLEDGMENTS

Wildlife Conservation International of the New York Zoological Society sponsored the research with additional financial assistance and language training provided by the Michigan State University African Studies Center through a U.S. Department of Education National Resource Fellowship. The study was affiliated with the National Museums of Kenya, Institute of Primate Research, under permission granted by the Office of the President (permit # OP. 13/001/17 C 24/9). I extend my gratitude to the East African Herbarium staff and especially Ann Robertson for their assistance with the nursery trials. Special thanks go to Cy Young for his help with the GIS analyses and to Bakari Mohammed Garise for his assistance in the nursery and throughout the field research.

REFERENCES

Anderson, B. W., and R. D. Ohmart. 1985. "Riparian Revegetation as a Mitigating Process in Stream and River Restoration." In J. A. Gore, ed., *The Restoration of Rivers and Streams*. Boston: Butterworth. 41-79.

Ashby, W. C. 1987. "Forests." In W. Jordan III, M. E. Gilpin, and J. D. Aber, eds., *Restoration Ecology: A Synthetic Approach to Ecological Research*. Cambridge: Cambridge University Press. 89-108.

Bradshaw, A. D. 1987. "The Reclamation of Derelict Land and the Ecology of Ecosystems." In W. Jordan III, M. E. Gilpin, and J. D. Aber, eds., *Restoration Ecology: A Synthetic Approach to Ecological Research*. Cambridge: Cambridge University Press. 53-74.

Burrough, P. A. 1986. *Principles of Geographical Information Systems for Land Resources Assessment*. Oxford: Clarendon Press.

Cairns, J., Jr. 1988. "Restoration Ecology: The New Frontier." In J. Cairns, Jr., ed., *Rehabilitating Damaged Ecosystems*. Boca Raton, Fl.: CRC Press. 1:1-11.

Decker, B. S. 1989. *Effects of Habitat Disturbance on the Behavioral Ecology and Demographics of the Tana River Red Colobus* (Colobus badius rufomitratus). Ph.D. dissertation, Emory University.

Decker, B. S., and M. F. Kinnaird. 1992. "Tana River Red Colobus and Crested Mangabey: Results of Recent Censuses." *American Journal of Primatology* 26: 47-52.

Eastman, J. R. 1990. *IDRISI. A Grid-Based Geographic Analysis System*. Version 3.2.2. Worcester, Mass.: Clark University, Graduate School of Geography.

FAO. 1984. *Agriclimatological Data for Africa*. Vol. 2, *Countries South of the Equator*. FAO Plant Protection Series No. 22. Rome: Food and Agriculture Organization of the United Nations.

Franklin, J. F., P. M. Frenzen, and F. J. Swanson. 1988. "Re-creation of Ecosystems at Mount St. Helens: Contrasts in Artificial and Natural Approaches." In J. Cairns, Jr., ed., *Rehabilitating Damaged Ecosystems*. Boca Raton, Fl.: CRC Press. 2:1-37.

Gilpin, M. E. 1987. "Spatial Structure and Population Vulnerability." In M. E. Soulé, ed., *Viable Populations for Conservation*. Cambridge: Cambridge University Press. 125-39.

Greig-Smith, P. 1983. *Quantitative Plant Ecology*. Berkeley: University of California Press.

Hamilton, A. 1974. "The History of Vegetation." In E. M. Lind and M. E. S. Morrison, eds., *East African Vegetation*. London: Longman Group Limited. 188-209.

Harris, L. D. 1984. *The Fragmented Forest*. Chicago: The University of Chicago Press.

Holdridge, L. R. 1967. *Life Zone Ecology*. San Jose, Costa Rica: Tropical Science Center.

Hughes, F. M. R. 1987. "Conflicting Uses for Forest Resources in the Lower Tana River Basin of Kenya." In D. Anderson and R. Grove, eds., *Conservation in Africa*. Cambridge: Cambridge University Press. 211-28.

———. 1988. "The Ecology of African Floodplain Forests in Semi-arid and Arid Zones: A Review." *Journal of Biogeography* 15: 127-40.

Kinnaird, M. F. 1990. *Behavioral and Demographic Responses to Habitat Change by the Tana River Crested Mangabey* (Cercocebus galeritus galeritus). Ph.D. dissertation, University of Florida.

Kinnaird, M. F., K. E. Medley, B. S. Decker, and W. O. Ochiago. 1990. *Management Issues and Recommendations for the Tana River National Primate Reserve, Kenya*. Report submitted to Wildlife Conservation International of the New York Zoological Society and the Kenya Wildlife Service, Nairobi, Kenya.

Ledec, G. 1987. "Effects of Kenya's Bura Settlement Project on Biological Diversity and Other Conservation Concerns." *Conservation Biology* 3 (1): 247-58.

Lovejoy, T. E., R. O. Bierregaard, Jr., A. B. Rylands, J. R. Malcolm, C. E. Quintela, L. H. Harper, K. S. Brown, Jr., A. H. Powell, G. V. N. Powell, H. O. R. Schubart, and M. B. Hays. 1986. "Edge and Other Effects of Isolation on Amazon Forest Fragments." In M. E. Soulé, ed., *Conservation Biology*. Sunderland, Mass.: Sinauer Associates, Inc. 257-85.

Lusigi, W. 1981. "New Approaches to Wildlife Conservation in Kenya." *Ambio* 10: 87-92.

McIntosh, R. P. 1985. *The Background of Ecology*. Cambridge: Cambridge University Press.

Marsh, C. W. 1976. *A Management Plan for the Tana River Game Reserve*. Report to the Kenya Department of Wildlife Conservation and Management. Bronx, N.Y.: New York Zoological Society.

———. 1986. "A Resurvey of Tana River Primates and Their Forest Habitat." *Primate Conservation* 7: 72-81.

Medley, K. E. 1990. *Forest Ecology and Conservation in the Tana River National Primate Reserve, Kenya*. Ph.D. dissertation, Michigan State University.

———. 1992. "Patterns of Forest Diversity along the Tana River, Kenya." *Journal of Tropical Ecology* 8: 353-71.

———. 1993. "Primate Conservation along the Tana River, Kenya: An Examination of the Forest Habitat." *Conservation Biology* 7: 109-21.

Medley, K. E., M. F. Kinnaird, and B. S. Decker. 1989. "A Survey of the Riverine Forests in the Wema/Hewani Vicinity with Reference to Development and the Preservation of Human Resources." *Utafiti* 2 (1): 1-6.

Muchena, F. N. 1987. *Soils and Irrigation of Three Areas in the Lower Tana Region, Kenya*. The Netherlands: University of Wageningen.

Newmark, W. D. 1987. "A Land-bridge Island Perspective on Mammalian Extinctions in Western North American Parks." *Nature* 325: 430-32.

Pickett, S. T. A., and J. N. Thompson. 1978. "Patch Dynamics and the Design of Nature Reserves." *Biological Conservation* 13: 27-37.

Shafer, C. L. 1990. *Nature Reserves: Island Theory and Conservation Practice*. Washington, D.C.: Smithsonian Institution Press.

Young, C. W., and W. H. Renwick. 1992. *IMAP User's Guide*. Version 1.3. Oxford, Ohio: Department of Geography, Miami University.

Remaking and Restoring the Landscape of Dare County, North Carolina

John E. Wierwille

From the Virginia capes, the North Carolina coast sweeps south in a series of graceful arcs. The outermost area is the preeminent feature of this shoreline: a thin ribbon of sand separated from the mainland by two large, shallow embayments, the Albemarle and Pamlico sounds (see fig. 1). Its physical components are called barrier islands because they buffer the high-energy waves and storm surges of the Atlantic, thereby protecting the mainland (Dolan and Lins 1987). The role these islands play in providing habitat is equally significant. The backsides of North Carolina's barrier islands are lined by tidal marshes. Here, an abundance of emergent vegetation produces the organic matter that begins the food web of this estuary system, the second largest in eastern North America. In addition to their role as producers, these islands also create beaches, dune systems, maritime forests, tidal salt marshes, and mixed and freshwater estuaries. Measured by the ecological standard of diversity, few systems compare.

What really makes this landscape unique are the dynamic geomorphic and violent atmospheric forces that shape it. The way a coastal system responds to the constant pressures presented by tidal and wave pressure is determinative of its physical configuration as well as its biological components. In areas as prone to the more dramatic influence of storms as coastal North Carolina, the ways in which the coast transfers tidal and wave energy are even more relevant. Barrier islands respond to these combined forces by simply migrating in the general direction in which they are pushed by the combination of wave, tidal, and storm-generated forces. In a process called "overwash," severe storms push sand from the

168

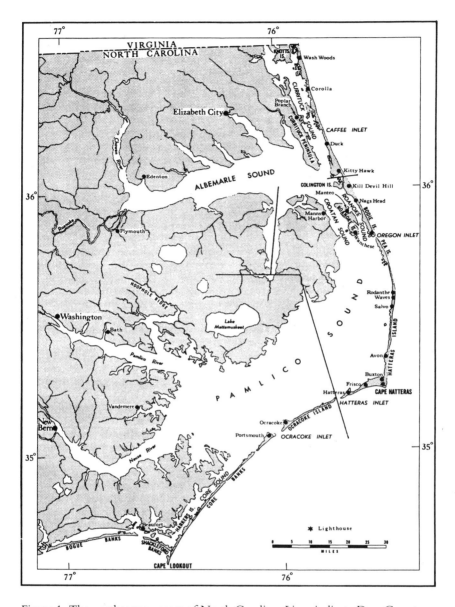

Figure 1. The northeastern coast of North Carolina. Lines indicate Dare County.

exposed side of islands over the low dune berm (Hayes 1979). The end result is a dramatic dissipation of the constant pressure presented by wave energy. Furthermore, the high salinity and recurring burial of overwashed areas ensures sparse vegetation on the beach and dunes, thereby inhibiting stabilization of dunes (Spitsbergen 1980). The low dunes contribute, in turn, by fostering overwash, which maintains the wide, flat shape of the islands. Finally, the rush of water forced into and from the two sounds by each storm can move enormous quantities of sand, regularly opening and closing inlets between islands. Since European settlement, for example, at least fifteen inlets have opened or closed on the Outer Banks (Dunbar 1958). In effect, the system succeeds by succumbing in every way to the violence it encounters. The combination of sparse vegetation, a flat berm of loose sand, and a wide profile encourages overwash. The position and shape of the islands are sacrificed, but the tidal marshes and estuarine waters are protected from the ravages of salinity and heavy waves. Shifting sands are the norm here, the price paid to protect the botanical and biological resources of the landscape.

These coastal ecosystems are not devoid of human influence. Indeed, the modern landscape is dominated more by social, cultural, and economic influences than by physical, chemical, and hydrologic forces. After four centuries of combating the wind and sea, early settlers and the generations that followed have temporarily subverted those forces and replaced the ecologic landscape with a purely commercial one. Huge portions of the Albemarle-Pamlico peninsula are farmed. Large fishing vessels ply dredged channels in the sounds and shelter in constructed harbors. A jetty temporarily prevents the closing of at least one inlet to provide access to ocean boat traffic. Cottages blanket the islands from beach to marsh. A network of roads, bridges, and ferries connects land areas and, more important, provides access to the affluent and growing populations of the Chesapeake Bay and is thus the life-blood of the area's tourist economy.

Human structures and patterns of land use are more than mere additions to this landscape; they have profound ecological functions as well. The resultant landscape is marked by conflict, and often by changes in the natural ecosystems. For example, agricultural residues and constant dredging influence estuarine chemistry enough to reduce habitat quality, and fixed structures on the islands encourage rapid land loss.

These conflicts have social and economic implications, too. Violent storms regularly strike this coast, threatening great property damage. Further complicating matters, the federal government insures most privately owned property on the islands (Kuehn 1981). Thus, storms not only create major physical changes in the landscape, but also impose major economic loss and an enormously expensive federal liability (Marx 1991).

The socioeconomic attitudes that have manipulated this landscape are especially revealing. Derived from an economic and cultural ethos rather than from ecological dictates, they result in seeming incompatibility of human and nonhuman constructs. Are the differences so great they cannot be reconciled? Does human use of coastal resources preclude ecologic health? Can the positive ecologic functions of barrier islands and estuaries be restored? Although such questions ultimately must be addressed on all the barrier islands that stretch southward from Massachusetts to Florida and then westward into the Gulf of Mexico, the Outer Banks of North Carolina are particularly well suited to reveal the issues of landscape restoration as championed by William Jordan in this book.

Dare County contains more barrier islands and more estuarine acreage than any other county in North Carolina, and is an optimal area of study for many reasons. First, its historical record extends for thousands of years, and includes 350 years of continuous occupation since European colonization. Second, several detailed ecological studies have examined both island and estuarine ecosystems there. Third, with over 150 years of resort history, this area is the key to coastal tourism in North Carolina. Lastly, the ecosystems of Dare County have been subjected to repeated interference and wide-scale manipulation by a variety of capital ventures. Thus, the county has been the site of an amazing struggle between humans and natural forces for control of an entire landscape. The history of this struggle reveals not only the changes that have occurred in the Dare landscape, but also the sources of conflict. More important, perhaps, it hints at changes that might occur in the future and suggests ways to avoid further conflict.

The Ecological Remaking of Dare County

Historian Alfred Crosby has argued that European colonization of the New World has been marked by efforts to remake landscapes as "Neo-Europes" (Crosby 1986). Such a remaking was undeniably a goal of the first Europeans to encounter the Outer Banks. John White, reporter for the first English venture to the area, depicted it as "England iff only we wishe it," implying with great clarity the landscape the colonists sought to create (Powell 1989). Almost four centuries later, the remaking continues, albeit guided by a different vision.

The first colonists had little impact on area ecosystems. Despite attempts at increasing commercial activity, they simply lacked the technology, capital, and means of transportation necessary to have more than superficial impacts on ecological systems. One exception was the cedar shingle export industry, which nearly extirpated that species from the area (Dunbar 1958). The introduction of free-ranging stock, however, represented a much more serious and unforeseen threat to the dune system. The increased soil disruption, defoliation by grazing, and the nitrogen these herds added had immediate impacts. It is clear that every effort was made to maximize the livestock reproduction. Evidence suggests herd size was not checked to prevent overgrazing. When an island's population of cattle became too dense, the owner simply removed the excess to market or allowed the elements to thin the numbers (Thomson 1942). Thus, what might have been only a minor pressure was exponentially compounded.

Almost certainly, returns on livestock and more commerce in the mid-Atlantic region were the impetus to increased settlement in the area, and as the colonials developed a familiarity with the land, the manipulations they proposed were of ever greater number and scale. The first Europeans dreamed of what the land might be; the colonials began the remaking. The culture was willing to change the environment if a profit would result. The limiting factor to this point was the absence of the necessary technology and capital. Even the legal system reflected this approach. Until the early 1700s, any free male could claim ownership of a parcel of land provided he "improve" its usefulness, generally a combination of clearing land for agriculture and introducing livestock (Cushing 1977).

By the early 1800s, capital became available for the first truly extractive ventures. Fishing in the sounds and rivers became an important industry, around which rose the settlements on the peninsula. Changes in the estuaries occurred almost immediately. The populations of several anadromous fish species were reduced dramatically by seasonal catches during their return to spawn (Dunbar 1958). The impacts on the islands increased by degree also, as livestock herds expanded and the first resorts were established. The latter heralded the most significant change for the islands, as settlement patterns shifted from the protected backside to the open, breezy, ocean beaches (Dunbar 1958). From this point forward, most human structures were built either on the beach berm or among the dunes. Thus, each cottage, road, and lighthouse constructed now obstructs the overwash process by stabilizing the dune structure. With overwash prohibited, the constant pounding from ocean forces has resulted in rapid land loss (Pilkey et al. 1978). Finally, hunting and timbering during the late colonial period disrupted the plant and animal communities. By the early 1800s, wolves became the first of many animal species to be extirpated from the area (Cushing 1977).

The mid-1800s were characterized by the first noted changes in the environment. Overgrazing nearly defoliated many of the islands so that unanchored sands were, for the first time, influenced by winds as well as the overwash process. Sometimes, the blowing sand even posed problems for humans, burying the few remaining woods and disrupting tourism (Ruffin 1861). Fortunately for island plant communities, economic pressure and preference for selected breeding forced the free-ranging livestock industry to fold before the islands were devoid of all vegetation (Thomson 1942). Meanwhile, the commercial fishing industry expanded and as it did, like the livestock industry before, its impact grew proportionally. Several species' populations dropped to unharvestable levels before the Civil War (Dunbar 1958). The unwitting push for profits continued to deliver environmental change.

After the Civil War, northern investors provided long-awaited technological inputs and capital investment. The results were disastrous. Shad, alewive, scallop, striped bass, oyster, terrapin turtle, menhaden, and clam fisheries all operated with success, and then failed because of overfishing in the forty years following the war. For these species, the impact was so great that populations have taken nearly a century to return to preharvest levels (*Natural Resources Bi-Weekly* 1925). In addition, shrimp and crab fisheries developed. Although these species have been periodically overfished, their populations have recovered to profit-sustaining levels in much shorter periods of time (*Conservation and Development Bi-Weekly* 1924). These were the first real and substantial influences of humans on the aquatic food web of the Carolina estuaries.

During the 1920s and 1930s, manipulations reached new extremes when island vegetation patterns were thoroughly altered. In the shallow estuaries, eel grass, the primary emergent in tidal marshes, was harvested so extensively for sale as hay that the entire estuarine system nearly collapsed (Stick 1970). On dry land, the New Deal of the 1930s initiated a "reconstruction" program for all the islands between Virginia and Cape Hatteras. Under this program, the Civilian Conservation Corps revegetated islands from beach berm to tidal flat with 142 million square feet of sod and more than 2.5 million seedlings and shrubs. In addition, the CCC erected more than 115 miles of sand fences to stabilize dunes and prevent encroachment of the sea (Stick 1970). Measured by the short-term goal of stabilizing the loose sands, the programs were declared a success. Indeed, the area was covered by vegetation where none had grown since the arrival of Europeans.

Appearances and reality can be two different worlds, however. The stabilization efforts of the CCC actually represent the beginning of the most dramatic and profound changes in the Outer Banks system to date. The effect of constructing stabilized barrier dunes was a fundamentally different island geomorphology. Fifty years later, comparison of altered and unaltered islands shows that stabilized islands experienced exponentially greater shoreline loss than unstabilized islands because of decreased overwash (Dolan et al. 1973). Furthermore, though more sparsely vegetated, the unaltered islands maintained the wide flat profile characteristic of unaltered barrier islands (Lins 1980). The move to stabilize the islands had begun, and with it humans began to subvert the collection of non-human forces so entwined with the ecologic health of the Outer Banks.

The CCC programs are also significant as they represent the first of several substantial federal investments in Dare County. Bridges, ferries, several roads, and a few harbors were built with federal funding at this time (Monaghan 1985). Only the federal government had the huge equipment, labor, and capital resources necessary to undertake such enormous projects. In subsequent years, Dare residents have often returned to this source of investment capital to help remake the landscape.

The complicity has not always been so direct. Tourism became the industry of choice for Dare County in the early 1950s. Just as waves had shaped the islands in the past, tourism became the wave of the future, overlaying the landscape with beach houses, motels, boardwalks, museums, aquariums, packed-in-so-tight-you-can't-move campgrounds, and gift shops with bags-of-shells and bottles-of-sand. Tourism brochures and local tourist bureau activity reveal a simple strategy of divided interests: local investment should work to attract more people and increase market share, while government contributes a secure, safe, and modern infrastructure (Brown 1950). In the late 1960s, the federal government began the National Flood Insurance Program (NFIP), providing inexpensive flood insurance for municipalities complying with federally mandated zoning regulation. Despite legislative intent, the NFIP acted as an indirect incentive to development. Thousands who could not afford the risk associated with building in the face of storms were now assured of financial security (Kuehn 1981). The 1980 census showed that the population had doubled and the number of structures tripled (Monaghan 1985).

Today, the dunes and beaches of the Cape Hatteras National Seashore seem more unnatural and out of place than the proliferation of roads, beach houses, hang gliders diving from an enormous dune, and air-conditioned aquariums. This is the other remaking that has occurred in Dare County. Physical, chemical, and biological landscapes are not only altered, they are removed from view by human commerce, buried beneath the foundations, roads, and bridges. The economy is no longer based on extracting resources; instead it is associated with the perception of a particular landscape. So long as tourists believe they are escaping to an ecologically healthy place, they will continue to return. The fact of the matter, of course, is that perception is not reality. The economy is still fundamentally tied to the health and vigor of supporting ecosystems, and the changes in that system are very real and tangible. As yet, the changes in the landscape have had little impact on tourism, but future development—offshore drilling, for example—might prove more disruptive than other modern enterprises.

The Future of Dare County

The future of Dare County is uncertain. In 1993, millions of tourists will visit, hundreds of new beach homes will be built, and expanded utility services will attract even more people to the area. The result will be greater demands on the county's supporting ecosystems. Dredging will maintain open passage through the sounds and Oregon Inlet, suspending tons of sediment in the waters of Pamlico and Albemarle sounds each day. Every year, thousands of gallons of herbicides and insecticides will run off into the sounds from the peninsulas. Estuarine chemistry will change with each new factor introduced. Biological systems will respond accordingly. Perhaps primary production and/or diversity of species will suffer as well. Some species might be extirpated or even become extinct. More likely, primary production will diminish, with populations of species that support human economies—game fish, shellfish, and other commercially attractive species—declining accordingly. These changes will be noticed. Fishing, tourist, and environmental interests probably will be advocated in response. Policymakers will have to try to balance these interests and plot a course for the future.

Undoubtedly, the benefits and costs will be weighed, but how value is determined will decide the issue. Tourism interests, land-use policies, and judicial decisions concerning landowner obligations are the human factors involved. Atmospheric events, biological processes, and geomorphic patterns will be the nonhuman influences. Despite many factors, the problem reduces to two questions:

1. To what extent will efforts to stabilize the islands, especially the beach zone, be continued?
2. What effect will the increased turbidity from dredging, decreased salinity from increased runoff, and influx of agricultural chemicals have on the food web of the estuarine sounds?

Any number of outcomes is possible. Beach stabilization could be expanded as in New England, or engineering efforts could be stopped entirely. Even the expensive option of "building" beaches through sand nutrification might be chosen. The impacts will certainly be more widespread and recognizable. Unfortunately, as has often been noted, public awareness generally occurs only after ecological dysfunction is irreversible.

These threats to the ecological systems of Dare County will be limited by federal, state, and local land-use policy. Current regulations preventing the drainage of wetlands provide some of the strongest environmental protection in the United States. Neither the peninsular nor tidal marshes are likely to face serious threats anytime soon. Further, a majority of remaining wetlands are owned and protected by federal or state government or private groups as wildlife refuges or as a national park. Threats to the beaches also face increased regulation. Amended in 1991, the federal Flood Insurance Protection Act now limits insurance coverage to existing structures, and disaster assistance is not available for municipalities that fail to enact zoning regulations to limit beach construction (U.S.C.A. 1991). This effort is complemented by the Coastal Area Management Act (CAMA) adopted by the North Carolina General Assembly. Written by state and local legislators according to federal land-use policy, CAMA is a contract between federal and state governments. For enacting and enforcing local policies that comply with federal policy goals, municipalities are assured monies and legal assistance (Brower and Carol 1991). Together, these policies will undoubtedly limit the damage posed by further unwise development.

Dare County's future depends on the values and priorities of humanity—those involved in policymaking and those who can effectively lobby their interests. The Dare landscape will be used for something. It will remain developed for hundreds of years. Human use could be aimed at more than economic returns, however. Policies can be chosen that endeavor to restore the landscape to a more natural state and establish Dare County as an example of the restoration of a once-damaged coastal ecosystem.

The Need for a New Land Ethic

Even if restoration is pursued, the salvation of Dare County will not be found solely in policy papers and legal decisions. The activities that shaped this landscape have been entirely within the bounds of law. In fact, the most significant manipulations have been designed and completed by government effort. Legal possession of the land has been translated to mean "sovereignty over the land." The inherent rights of owners under this system are at the root of the environmental problems in Dare. Aldo Leopold (1949) recognized this decades ago:

> There is as yet no ethic dealing with man's relation to the land and to the animals and plants which grow upon it. Land . . . is still property. The land-relation is still strictly economic, entailing privileges but not obligations. (203)

In the United States, property rights are a fundamental foundation on which our legal system has been constructed. Because of this, however, the legal system tends to limit the linkage of obligations to property ownership. As Donald Worster (1979) has noted, the result is that "privileges are continually derived, but never repaid. The landscape is abused by a few for the profit of a few, but healed on the grounds of common ownership at the expense of all citizens" (5-7). With the loss of each species, mile of shoreline, and acre of wetland, it becomes abundantly clear that the scale and type of economy this set of relations advances are not conducive to ecological equilibrium.

Dare County's ecosystems have been disrupted largely by introduced, large-scale technologies. The disruption began at a time when modernity had convinced humanity that all things were possible; even the incredible forces of the ocean could be turned away. A lesson it did not teach is that all things have a price. Leopold (1949) also recognized the disparity between his land ethic and the conquering fixation of humanity:

> A land ethic changes the role of Homo sapiens from conqueror of the land-community to plain member and citizen of it. It implies respect for his fellow-members, and also respect for the community as such.

> In human history, we have learned (I hope) that the conqueror role is
> eventually self-defeating. Why? Because it is implicit in such a role that
> the conqueror knows, ex cathedra, just what makes the community
> clock tick, and just what and who is valuable, and what and who is
> worthless, in community life. It always turns out that he knows neither,
> and this is why his conquests eventually defeat themselves. (204)

Complete understanding of any environment is impossible. Dare County is no exception, but we are just beginning to understand the significant role of this type of coastal system. Barrier islands may prevent billions of dollars in property damages each year. They provide recreation for nearly one-third of the U.S. population each year (Marx 1991). The tidal marshes they create are perhaps the most productive ecosystems in the world, generating more than ten tons of useful organic material per acre—more than twice the yield of a cornfield, ten times as much as coastal waters, and thirty times as much as the open ocean (Berger 1979). As such, they are a compelling example of the need to embrace biocentrism and discard anthropocentricity.

Now that the beneficial aspects of this system—shoreline protection, estuarine resources, and recreational potential—have become more apparent, the reasons for past incompatibility have also become more clear. Consider the efforts of the CCC during the 1930s, for example. Two inadequacies generally explain that failure. First, the special role of the dynamic barrier-island geomorphology in determining the ecological balance of the estuarine system was simply not known. Overwash itself had only been speculated at that time; no one suggested it might be so important. Second, the effort was conceived and performed from an entirely anthropocentric perspective. The human problem was constantly shifting sands. The best answer seemed to be stabilization. The combination led to a program designed in direct opposition to the oceanic and atmospheric forces that dominate this system.

To the credit of the CCC, it should be noted that only native grasses and shrubs were chosen for its planting projects. Perhaps that decision stemmed from necessity: very few nonnative species can tolerate the extreme soil deprivation and constant overwash to which the native plants are adapted. Perhaps it was intentional; the record does not show. This sort of "concession," however, is the key lesson to be taken from past failures. Humans must learn to work with the materials at hand and within the scope of the ecosystem they choose.

The Outer Banks are not going to disappear. They will hug the North Carolina coast for centuries to come. Humans will also be a part of that landscape for centuries. The challenge is to maintain the ecological diversity of species and the special geomorphology during those years of use. It seems entirely possible, with everything that is now known, to accomplish that balance. Over the course of years, construction should move to the backside of the islands. The stabilizing features of human development—roads, pipelines, jetties, seawalls, and bridges—should be avoided if at all possible or alternative locations found. Finally, both policymakers and users must remember that they are merely a part of this ecosystem, not its end. If only those three guides are followed, the system will likely maintain its millennia-old character.

The shifting sands of the Outer Banks may be the bane of human development and fairly impervious to repeated assault, but they need to be respected and valued as a component of the biosphere. Neither the relative absence of life there nor the value they represent as real estate reduces the ecological worth of these islands and their geomorphic processes. In fact, perhaps because they *are* so volatile and unaccepting of permanent habitation, a balanced state of use based on biocentric values can be found that still considers the social and economic value of the islands. Restoration ecologists seek precisely this. As William Jordan (1991) describes it, the goal is to learn to fit our own human systems into the nonhuman systems more closely.

> The important thing is to pursue our experiment hopefully, in the
> expectation of finding every element of harmony we can in the
> relationship between ourselves and the rest of nature, rediscovering old
> harmonies and perhaps finding some new ones too. (103-7)

This is the option I would advocate for Dare County: the vigorous pursuit of policies—relationships with the land—that choose well-placed efforts that contribute to the diversity of surroundings, restoring habitats rather than further degrading them.

REFERENCES

Berger, John. 1979. *Restoring the Earth.* New York: Alfred A. Knopf.

Brower, David J., and Daniel S. Carol. 1991. *Coastal Zone Management as Land Planning.* Washington, D.C.: The National Planning Association.

Brown, Aycock. 1950. "Dare Ready for Tourists." *The Raleigh News and Observer* (February 23): D1.

Conservation and Development Bi-Weekly. 1924. "Changing Attitudes in North Carolina Fisheries." 11, 16 (November 29): 3-4.

Crosby, Alfred. 1986. *Ecological Imperialism: The Biological Expansion of Europe, 900-1900.* New York: Cambridge University Press.

Cushing, John. 1977. *Earliest Printed Laws of North Carolina.* Wilmington, Del.: Michael Glazier, Inc.

Dolan, Robert, and Harry Lins. 1987. "Beaches and Barrier Islands." *Scientific American* 257 (July): 68-71.

Dolan, Robert, Paul J. Godfrey, and William E. Odum. 1973. "Man's Impact on the Barrier Islands of North Carolina: A Case Study of the Implications of Large-Scale Manipulations of the Natural Environment." *American Scientist* 61 (March-April): 152-62.

Dunbar, Gary. 1958. *Historical Geography of the North Carolina Outer Banks.* Baton Rouge: Louisiana State University Press.

Hayes, Miles O. 1979. "Barrier Island Morphology as a Function of Tidal and Wave Action." *Barrier Islands from the Gulf of St. Lawrence to the Gulf of Mexico.* New York: Academic Press. 1-27.

Jordan, William R. 1991. "Editor's Comments." *Restoration & Management Notes* 9 (Summer): 2.

Kuehn, Robert A. 1981. "The Shifting Sands of Federal Barrier Island Policy." *Harvard Environmental Law Review* 5 (Summer): 217-58.

Leopold, Aldo. 1949. *A Sand County Almanac; And Sketches Here and There.* New York: Oxford University Press.

Lins, Harry. 1980. *Patterns and Trends of Land-Use and Land-Cover on Atlantic Coast Barrier Islands.* Washington, D.C.: U.S. Government Printing Office.

Marx, Wesley. 1991. *The Frail Ocean: A Blueprint for Change in the 1990s and Beyond.* Chester, Conn.: Globe Pequot Press.

Monaghan, George. 1985. *Community Planning for the Future.* Raleigh: Division of Community Planning, North Carolina Department of Conservation and Development.

Natural Resources Bi-Weekly. 1925. "Shall State Marsh Lands Be Lost to So Many for So Few?" 2, 19 (January 10): 1-4.

Pilkey, Jr., Orrin H., William J. Neal, and Orrin H. Pilkey, Sr. 1978. *From Currituck to Calabash: Living with North Carolina's Barrier Islands.* Research Triangle Park, N.C.: North Carolina Science and Technology Research Center.

Powell, William S. 1989. *North Carolina through Four Centuries.* Chapel Hill, N.C.: University of North Carolina Press.

Ruffin, Edmund. 1861. *Agricultural, Geological, and Descriptive Sketches of North Carolina and the Similar Adjacent Lands.* Raleigh: Institution for the Deaf & Dumb & Blind.

Spitsbergen, Judith M. 1980. *Seacoast Life: An Ecological Guide to Natural Seashore Communities in North Carolina.* Raleigh: North Carolina Museum of Natural History.

Stick, David. 1970. *Dare County: A History.* Raleigh: North Carolina Department of Cultural Resources, Division of Archives and History.

Thomson, J. W. 1942. *A History of Livestock Raising in the United States, 1607-1860.* Washington, D.C.: USDA, Bureau of Agricultural Economics.

U.S.C.A. 1991. *Amendments to the National Flood Insurance Act.* Sections 1489-1497.

Worster, Donald. 1979. *Dust Bowl: The Southern Plains in the 1930s.* New York: Oxford University Press.

Rehabilitation of Land Stripped for Coal in Ohio—Reclamation, Restoration, or Creation?

A. Dwight Baldwin, Jr.

> Wealth and power were built, and to a large extent rest, on the exploitation and industrial use of mineral resources. This is a process which inevitably makes a mess of the land. Our forebears, for the most part, left the mess as it was: either they did not mind it, or they found the task of cleaning it up too difficult and too costly.
>
> Civic Trust 1964

> It is not too late to repair some of the mistakes of the past and to make America a green and pleasant—and productive—land. We can do it if we understand the history of our husbandry, and develop fresh insight concerning the men and the forces that have shaped our land attitudes and determine the pattern of land use in the United States.
>
> Stewart L. Udall, *The Quiet Crisis*

Surface mining of mineral resources across the United States has left a legacy of highly disturbed landscapes, many of which are sources of aesthetic, physical, and chemical contamination. Through the impetus of both state and federal legislation, however, much has been learned about the steps necessary to reclaim such land for productive use and to minimize environmental degradation. It should be emphasized that few if any mining companies would claim that they restore the landscape to conditions that existed prior to mining; true restoration would be impossible given our limited knowledge of the finer workings of ecosystem dynamics. In addition, state and federal mining regulations declare that the land must be reclaimed to "higher and better use" unless the premined land was already prime farmland, in which case every attempt must be made to return the land to its previous condition, i.e., farmland. Although this legal mandate requires land restoration, most would agree that even in this exceptional case soil properties such as porosity and permeability are not the same as those found in the field prior to mining.

Thus the issue of reclamation to "higher and better use" has engendered considerable debate because mined land reclaimed for higher and better use in Ohio does not in any way resemble the human-made "healthy and eco-logically rich landscapes" championed by Frederick Turner in his lead paper for this book. This chapter addresses the reasons for this disparity. After a description of reclamation practices employed by coal-mining companies in Ohio, three questions provide a focus for the second portion of the chapter: (1) What considerations have governed decisions as to the best use of the land after mining? (2) Have those decisions served the best interests of the people of Ohio? and (3) Should the objective of mine-land rehabilitation be reclamation, restoration, or creation?

Regional History

Prior to 1800, eastern Ohio was heavily forested by mature hardwoods with little indication of human impact from either Native Americans or early set-tlers. By the 1850s, however, transportation routes were improving and the influx of European settlers brought about increasingly rapid environmental deterioration. Particularly devastating was the development of a fledgling iron industry in southeastern Ohio, which required large quantities of char-coal as a source of heat and as a reducing agent for iron oxides in the blast furnace. Vast tracts of forested land were cut to satisfy this demand, and in-creased erosion rates and river flooding resulted.

With this rapid exploitation of the old-growth forests came the real-ization that this fledgling industry would fail if a new source of carbon was not found; this source was soon discovered at shallow depths as coal. The first reported tonnage of coal mined in Ohio occurred in 1800, three years before Ohio achieved statehood. All of Ohio's production was from underground mines until the beginning of the First World War (Board on Unreclaimed Strip Mined Lands 1990).

The onset of surface mining in Ohio began with the shipment of large earth-moving equipment from Panama to Tuscarawas County in 1914 following completion of the Panama Canal (Charles C. King 1991, per-sonal communication). The introduction of this large equipment made it far cheaper and a great deal safer to mine coal by first stripping the soil and bedrock overlying the coal seams. By 1990, 61 percent of all coal mined in Ohio was produced by surface stripping, and Ohio ranked ninth in the country in the tonnage of coal produced (Weisgarber 1990).

Environmental Degradation Caused by Strip-Mining

Strip-mining for coal in Ohio involves the removal of all soil and bedrock above the coal seam. This waste material, called spoil, is then stacked in long windrows in areas where the coal has previously been extracted. Spoil is an appropriate term, for these piles of very permeable fragments of sandstone and shale can spoil the environment for many generations to come.

For example, erosion rates from land that has not been reclaimed are the highest of any land in Ohio. Sediment loss in amounts as high as two hundred tons per acre per year have been measured from abandoned mined land in eastern Ohio. In comparison, soil loss from fields planted in corn in Missouri was considered high when it exceeded nineteen tons per acre per year. The average erosion rate from most Ohio watersheds affected by past strip-mining was in excess of ten tons per acre per year (Soil Conservation Service 1985). In addition, an estimated thirty-six thousand acres of abandoned strip-mined land are producing sediment at a rate equal to or exceeding twenty-five tons per year. This rate is of such magnitude that this land will never develop a natural vegetal cover (Soil Conservation Service 1985).

Sediment erosion from spoil piles also results in the deposition of sediment in stream channels and on adjacent floodplains. Such sediment chokes the stream channel, thereby increasing the frequency and severity of flooding, and sediment deposition on the floodplain significantly decreases the agricultural productivity of the land. Approximately 3,435 acres of land have been covered by sediment from areas stripped for coal production (Soil Conservation Service 1985).

Spoil piles are also both chemically and physically unstable. The oxidation of the iron disulfide and weathering of the shales and sandstones in the spoil piles pollute 30 percent of the 4,589 miles of streams in eastern Ohio (Soil Conservation Service 1985), and many structures in eastern Ohio are threatened by landslides from stripped areas.

Finally, the barren and scarred landscape characteristic of abandoned mined land is a tremendous aesthetic and economic detriment to local communities and the region. The presence of such land lowers the value of adjacent property and restricts the potential of the region for residential, commercial, or recreational use.

Mining Legislation in Ohio

The first comprehensive strip-mine law in Ohio was passed in 1972. At that time, this act was one of the toughest in the country. Under the law and its subsequent amendments, an operator must apply to the Division of Reclamation, Ohio Department of Natural Resources, for a permit to mine and reclaim a specific site. By law the operator is obliged to (1) isolate and save the topsoil at the site before mining begins; (2) contour the mined area to a configuration approximating that of the original ground surface; (3) replace the topsoil over the reclaimed area; and (4) leave the site in a condition of equal or greater economic value than existed prior to mining. The operator must show that water pollution, flooding, erosion, landslides, and sedimentation downstream of the mine site will be prevented. To pay for the administrative cost of enforcing the law and to assure that the required work is accomplished, coal operators must pay a one-time acreage fee of $50 per acre as well as a $2,500 per acre bond fee. In addition, a severance tax on all minerals mined in Ohio is collected for use in reclamation of abandoned mined land and to plug abandoned oil and gas wells.

The federal Surface Mining Control and Reclamation Act (SMCRA) was passed in 1977, and in many ways it was fashioned after the 1972 Ohio legislation. Beyond spelling out reclamation requirements that must be followed by all coal operators in the country, a key element in this law was the severance tax of 35 cents per ton imposed on surface coal-mine operators for reclamation of abandoned mined land. Although not all the funds that are collected under this federal act are returned to Ohio, the state did receive $6,692,171 in 1989. This money, along with the severance tax of 9 cents per ton collected by the state, provides Ohio an average of about $7.3 million per year for the reclamation of abandoned mined land (Board on Unreclaimed Strip Mined Lands 1990).

Reclamation of Land Stripped for Coal

Neither the Ohio strip-mining code (Ohio Administrative Code Rule

1501) nor the federal Surface Mining Control and Reclamation Act makes reference to land restoration. Rather, they refer only to land reclamation. The differentiation is based on the fact that the destruction of landscape created by stripping coal creates such a major disturbance of premining ecosystems that it is economically and probably technically impossible to restore these sites to premining conditions. Thus reclamation as stipulated by law implies only that the land be returned to a "higher or better use" than existed prior to mining. The phrase "higher or better use" is defined as meaning that the land must have a higher economic value or a nonmonetary benefit after mining than was possessed by the land prior to the removal of the coal. This distinction between restoration and reclamation is important. If true restoration is not the goal, then many different options regarding final landscape design become possible. How are these decisions made and by whom?

There are basically two major categories of land reclamation in Ohio where such decisions must be made. The first is the reclamation of all land stripped since passage of Ohio's comprehensive strip-mine law in 1972. The second is land mined prior to 1972, which was seldom reclaimed because of ineffectual or nonexistent state and federal regulations. The latter poses the greatest environmental threat to Ohio and neighboring coal-producing states, for this orphaned landscape has been left scarred, unproductive, and a source of severe water contamination for many decades.

Active Mined-Land Reclamation

Before mining can commence, a coal operator must file a detailed plan that satisfies the Ohio Administrative Code Rule 1501 regarding how the coal will be mined and the land reclaimed. The code contains explicit regulations about the stockpiling of surface soils prior to mining, grading, spreading of the stockpiled soil, and planting of vegetation.

Of greater interest with respect to the theme of this volume than the details of the code is the rationale for and techniques used in reclaiming mined land. Although the Division of Reclamation must approve all plans for use of the land after reclamation, clearly the mine operator plays a key role in this decision process. The available options for land use after reclamation are: (1) cropland, including row crops, small grain, hay crops, nursery, and orchards; (2) pastureland occasionally cut for hay but used primarily for the grazing of livestock; (3) industrial parks, retail trade, or single- and multiple-family housing; (4) public or private lei-sure-time use; (5) forestland and fish and wildlife habitat; (6) storing of water for stock ponds, irrigation, and other such uses; and (7) undevel-oped land not managed in any fashion. As mentioned previously, the law requires that the land be reclaimed to cropland if it has been used for cropland for five or more years out of the last ten years prior to mining and is considered prime farmland.

Historically, the return of strip-mined land to "higher or better use" has been interpreted by the federal Office of Surface Mining Reclamation and Enforcement to mean restoration to cropland or pastureland. The heavy emphasis on the planting of herbaceous plants dominated by fes-cue and crown vetch in the 1960s and 1970s reflected the belief that this was the most effective way to control soil erosion. The planting of trees was discouraged because it was believed to be very difficult to establish trees in the compacted, acid soils that exist after grading of the spoil. Al-though the species of grasses and legumes that are typically planted today are more diverse, the primary vegetation cover preferred by the Ohio Di-vision of Reclamation is still herbaceous plants. This selection is the op-tion of choice because it is the quickest and most economical method of stabilizing the barren-erodible spoil after mining (J. Backs 1992, personal communication).

The emphasis on the return of mined land to crop/pastureland has also been encouraged because of strict survival standards set for land that is to be reclaimed to forest. The Ohio code stipulates that a stocking rate of 600 trees per acre must be planted after completion of grading of the site. Of these 600 trees, 450 countable trees per acre must be growing at the end of the five-year bond period for land. Of these 450 trees, at least 80 percent must have been in place for at least three years. Adding to the problem of establishing trees on reclaimed land is the need to keep her-baceous ground cover away from each tree to minimize competition for water and nutrients. Not only does this require time and effort and there-fore money, but it exposes bare spoil to erosion, which may violate ero-sion provisions stipulated in the mining permit. Many coal operators thus do not consider forest as a postmining use of the land because of the difficulty of meeting reclamation standards. New techniques for enhanc-ing survival rates for trees in spoil are clearly needed. Some such methods are now being used in reclaiming abandoned mined land (discussed later in this essay), but they have not been widely adapted in the reclamation of recently mined lands.

Also leading to the decision to plant grasses and legumes instead of trees is the fact that most operators are considering the "higher or better use" over the short term. In comparison to the quick return realized by planting herbaceous grasses and legumes in just one growing season, the benefits of trees and shrubs are not realized until years after the legal responsibility for vegetation restoration has passed.

Thus much reclaimed land in Ohio consists of monotonous expanses of pastureland with little biodiversity in either plant or animal populations. Indeed, during the first five years after mining and before the coal operator receives all bond money paid to assure proper restoration, the operator must cut down any woody vegetation that seeds itself in the crop or pastureland. This practice mitigates against an ecologically diverse and aesthetically pleasing landscape that would include small groupings of trees interspersed with open fields.

More consideration should be given to the long-term social and economic benefits of land reclamation for wildlife habitat, recreation, and forest products when decisions are made as to what will be the "higher or better use" of the land. In some cases this will require the development of new ways of controlling erosion over the short term until the trees establish themselves, of eliminating competition of annuals around the base of the young seedlings, and of assuring that the soil is not too compacted by heavy earth-moving equipment. In other instances the techniques already exist and only await imaginative implementation.

Another practice that rules against the development of a diverse landscape is the requirement that the land be graded back to conditions that approximate the original contour of the premined landscape. This means that the highwall, the vertical wall of exposed overburden and coal left after the last shovel of coal has been removed, must be backfilled. Yet these last deep depressions, if located in relatively low topographic areas, often fill with unpolluted ground water and have the potential of producing excellent aquatic habitat if developed properly. Only now is the Ohio Division of Reclamation seriously considering allowing highwall remnants to remain after reclamation to permit the establishment of wetlands (J. Backs 1992, personal communication). Indeed, although the requirement to return the mined land to premined contours may have had a beneficial impact at the time the SMCRA was passed, Doll (1988) believes that reclamation technology has advanced to the point today that this requirement should be abolished entirely.

The use of wetlands as natural filters for the removal of suspended sediments as well as dissolved iron and manganese and for pH control is also meeting considerable resistance from the federal Office of Surface Mining Reclamation and Enforcement. Such wetlands, if constructed properly, could cleanse acid mine drainage of high concentrations of iron and manganese and raise pH levels to a point that they support prized fish and wildlife habitats. The success of such natural filtration systems has been shown in a wetland constructed by the American Electric Power's Fuel Supply Department in Coshocton County. Despite the documented success of this wetland in controlling acid mine drainage, the federal Office of Surface Mining Reclamation and Enforcement has not approved the technique because of concerns about the long-term effectiveness of the wetlands to control this problem (J. Backs 1992, personal communication).

Abandoned Mined Land Reclamation

An estimated 210,000 acres of land in Ohio mined prior to the passage of the Ohio strip-mine law in 1972 was left scarred and useless (Board on Unreclaimed Strip Mined Lands 1988). At a reclamation cost of between $10,000 to $20,000 per acre, the National Abandoned Mined Lands Inventory has calculated that $194 million is needed in Ohio alone to eliminate this problem (Board on Unreclaimed Strip Mined Lands 1990).

Funds for specific projects are acquired by the Ohio Division of Reclamation by the submission of proposals to the federal Office of Surface Mining Reclamation and Enforcement and from a severance tax of 9 cents per ton collected by the state. Because federal guidelines stipulate that federal money must first be spent reclaiming abandoned lands that pose public health and safety concerns and because of the large number of these first-priority sites in Ohio, no federal money has been spent for reclamation of environmentally sensitive areas.

Despite the restrictive federal guidelines, some of the most exciting and innovative attempts at mine-land reclamation are being performed on abandoned mined land using dollars generated by the Ohio severance tax. For example, the Ohio Division of Reclamation has planted 1.75 million tree seedlings on 1,240 acres during the past eight years. Of these, approximately one-third (650,000 seedlings) were planted in the 1989-90 biennium. Particularly impressive has been the development of nursery beds at Ohio's Marietta State Nursery for production of both hardwood and softwood seedlings inoculated with the fungus *Pitholithus tinctorius*. The association of the fungus and the seedling roots increases water and nutrient uptake, thus making the seedling more tolerant to drought, high soil temperatures, and the low pH conditions (Board on Unreclaimed Strip Mined Lands 1990). These inoculated seedlings are not now available to coal operators but as demand increases it is anticipated that private nurseries will begin meeting the need.

Other new methods of increasing the productivity of abandoned mined land include the addition of soil amendments to increase the organic content, nutrient concentrations, and soil porosity and permeability. Materials used have included composted municipal waste, sewage sludge, and mushroom compost. Each technique has its drawbacks, however, and it is clear that no single remedy will solve the problem of soil infertility.

Finally, some abandoned mined land in Ohio has been reclaimed at no cost to the state by giving approval to operators who have permits in adjacent areas to go back into an area and remine coal seams not removed by the first operator. Under the permit to remine abandoned mined land, the operator must reclaim the land to a higher use than existed in the unreclaimed condition; however, the operator is not required to improve the quality of water leaving the site. Thus, following reclamation, iron, manganese, and pH levels in drainage water must be no worse than existed prior to remining. In most cases, suspended sediment erosion rates are reduced and the land is returned to some sort of productive use.

Reflections

The reclamation of land stripped for coal has and is creating entirely different landscapes than existed in premining time in eastern Ohio. A high percentage of this land is being converted from forest rich in species diversity to cropland or pastureland, and the rationale for this conversion appears to be based on short-term economics and shortsighted restoration requirements. Such conversion meets the legal criteria of converting the land to a "higher or better use," and pasturelands are the most easily planted and maintained over the first five to ten years following mining. In addition, quick-growing grasses and legumes provide a rapid stabilization of the landscape, thereby reducing soil erosion from the mine site within the first growing season.

Yet the monotony of the reclaimed grassland has limited aesthetic value and greatly restricts biodiversity. Pasturelands require ongoing management and thus expense if they are to be maintained as open space.

Frederick Turner would argue that the state and federal mining regulations, and the regulators who blindly follow the easy and quick solution to serious erosion problems that exist immediately after mining, lack the vision to see the creative potential that strip-mined land offers. The precious opportunity to create an entirely new, healthy, and ecologically rich landscape has been wasted. He would say that we have not used our innate "sense of beauty [that] tells us what is relevant, what is likely, what is proper, what is fruitful." Rather we have opted for homeostasis—a condition whereby a numbing sameness to the landscape has been accomplished by converting countless acres of mined land to grassland. This "idea of 'sustainability' and general homeostasis is a profoundly unnatural goal" (see Turner, "The Invented Landscape," this volume).

If Turner's ideals are to be achieved in the coal country of eastern Ohio, newly developed reclamation techniques must in some cases be implemented and in others new methodologies must be developed. For example, technology that will assure a better survival and growth rate for a wide diversity of tree species while at the same time limiting soil erosion during the first years of growth must be found. Greater creativity in landscape design must be achieved by discarding the regulation that requires that land be returned to premined contour and by conscientiously working to create a diverse planting of trees, shrubs, and open fields.

Because of the vast area of abandoned mine land in eastern Ohio, these scars will remain as part of the Ohio landscape for years to come. This legacy will continue to pollute surface water and ground water, cause the destruction of prime farmland along river floodplains, and threaten engineered structures downslope from the mine areas. Reparation of the worst sites will cost a minimum of $194 million just to eliminate the adverse effects on the health, safety, and general welfare of the public (Board on Unreclaimed Strip Mined Lands 1990). This huge expenditure does not even begin to address the need for restoration of land that, although not a threat to public health, safety, and general welfare, continues to pollute surface water and ground water while bringing about a deterioration of neighboring land and aesthetic values. New sources of funding will have to be found to address these problems.

Because of the scarcity of funds, it is particularly incumbent upon us as a society to reach a consensus as to what truly represents the highest and best use of the reclaimed land. I would agree with Turner that we should be turning our energies to developing and implementing reclamation technologies that will permit the creation of entirely new and diverse landscapes incorporating woodlands, open meadows, and ponds. Much of the research that would permit such landscape conversion has already been done. All that is needed is a change of vision! We should be focusing our attention on developing harmonious landscape design to provide aesthetically pleasing, recreationally attractive, and economically rewarding uses for this disturbed land.

Would such rehabilitation of mined land accurately be termed "restoration" in the sense implied by William Jordan in his lead essay in this book? The answer is a resounding no, for it is difficult to argue that much of this landscape should or even could be returned to conditions that existed prior to stripping of the coal. Nor should such careful landscape planning be called "reclamation," for this descriptive term has inherited many negative connotations over the years. Rather, such rehabilitation would best be termed "creation" in the same ideal that Chance and the Sibyl, Hermione, created an entirely new and beautiful landscape on Mars in Turner's epic poem *Genesis*. Such a planned effort, involving an interdisciplinary team of botanists, zoologists, hydrologists, and soil scientists, could produce a landscape far richer in ecological diversity than anything that existed prior to coal extraction.

REFERENCES

Board on Unreclaimed Strip Mined Lands. 1988. *Biennial Report, 1986-1987*. Columbus: Ohio Department of Natural Resources. 1-33.

———. 1990. *Biennial Report, 1988-1989*. Columbus: Ohio Department of Natural Resources. 1-16.

Civic Trust. 1964. *Derelict Land: A Study of Industrial Dereliction and How It May Be Redeemed*. London: Civic Trust. 1-70.

Doll, Eugene C. 1988. "Relation of Public Policy to Reclamation Goals and Responsibilities." In John Cairns, Jr., ed., *Reclamation of Surface-Mined Lands*. Boca Raton, Fl.: CRC Press. 1:1-219.

Soil Conservation Service. 1985. *Assessment and Treatment of Areas in Ohio Impacted by Abandoned Mines*. Washington, D.C.: U.S. Department of Agriculture. 1-73.

Udall, Stewart L. 1963. *The Quiet Crisis*. Salt Lake City: Peregrine Smith Books. 1-73.

Weisgarber, Sherry L. 1990. *1990 Report on Ohio Mineral Industries*. Columbus: Ohio Department of Natural Resources, Division of Geological Survey. 1-142.

PART IV

IMPLICATIONS AND CONSEQUENCES

Concrete Beasts/Plaster Gardens

Ellen Price

"He tells me that he likes to think of his camera as a gun. When a young man, he both painted and hunted and given great intuition, a form of Zen, his pheasant came to meet his shot. But now he approaches the dead brace of birds and he asks: What is their existence?" Thus, Julien Levy, in the preface to a book of drawings by Henri Cartier Bresson, describes the photographer who at seventy years of age "decided to put down his camera and take up drawing again. As a photographer, he had lived by hunting and gathering, and by taking large risks, now he decided to live by slow and patient farming" (Bresson 1989, 11, 15).

We may be, as Frederick Turner suggests in "The Invented Landscape" (this volume), another species that is shaping our environment as other natural phenomena have. Pouring concrete into molds of animal forms, painting them, and putting them on lawns is not shaping living species or the ecosystems they occupy. The decorative reconstruction of animal forms, however, though requiring a different kind of investment and effort, is in some ways an analogue to the restoration of living plants and animals advocated in inventionist ecology. The conviction that we may re-create through scientific restoration or genetic technology whatever aspects of nature we need for scientific purposes is mirrored in the commonplace sight of concrete beasts. Many individuals may be relatively unfamiliar with prairie reconstruction or other examples of inventionist ecology, but one type of reconstruction they probably have seen are the concrete beasts that pose on front lawns. The process of making artwork based on these scenes has led me, like Bresson contemplating the brace of birds, to ask: What is their existence? As we think about the claims for restoration and preservation, we will need to try to answer that question.

The process of drawing, like the confrontation with the brace of birds, leads us to consider different aspects of a subject. First of all, drawing implies looking and seeing, and then translating visual information to a language of shape, mark, and line that takes form on a two-dimensional surface. But drawing entails both observation of physical form and reflections on the significance of the subject depicted. Through a highly individualized process of perception and translation, the pictorial content expands to take on other meanings. Thus, the documentation of a specific scene may be the starting point for a body of work that could depart from the initial subject into other areas of inquiry.

Art in our culture extends beyond what we encounter in museums or galleries to include decorative images and objects that we experience in everyday life. It is useful to note that the art and artifacts of so-called non-Western or nonindustrialized societies are often utilitarian objects, art objects that have a place in the daily routine. A wide range of consumer goods fits this description, from knickknacks for display in homes and outdoors to illustrations on cards and posters to patterns on housewares and linens. Because they touch so many people's lives and are so visible, these objects deserve examination and may well reveal more about the culture producing them than does traditional "fine" art.

As a visual artist, I am sensitive to the importance of the objects that we make and design as well as the everyday sights we see. The impulse to draw or make prints based on fairly common subjects springs from the desire to communicate a sense of their significance and to heighten the viewer's awareness of them as reflections of a specific culture. Lawn art, which has long been a subject of my work, is of particular interest because of the public nature of the display and the extent to which it symbolizes attitudes toward nature. Cultural artifacts such as these permeate our lives and reflect widely held opinions and attitudes (for example, that animals are cute); these artifacts continue by their very ubiquity to shape our opinions and attitudes about the environment, our roles within it, our responsibility to it.

As I look over the pictorial subject matter in a body of work that I began approximately ten years ago, I find clear connections with the key questions raised in this book: the motives, means, and consequences of reconstructing the forms of living plants and animals. The reconstruction or reinvention that has provided the pictorial content for my artwork can be found in scenes of domestic animals, in concrete lawn statuary, in plant nurseries, and, most recently, in displays in a zoology museum. In "The Invented Landscape," Turner views human civilization as "not only the protector and preserver but also as its restorer, propagator, and creator." I find these roles reflected on a small scale in the reconstruction of living animals in the form of concrete statuary. These artifacts reflect the desire for a connection with nature through the symbolic re-creation of wildlife that has been threatened or driven away by growths in population and by housing development.

I have been most directly involved with the "concrete beasts" of suburban landscapes. In this chapter I will discuss the experiences and thoughts that led me to consider the relationship between humans and the nonhuman natural world. I grew up in Queens, New York, and had little experience with rural or less-populous communities. Leaving the city was nearly always associated with vacationing, and the dichotomy between the city for everyday living and the country for enjoying unspoiled nature was firmly ingrained in me. I formed a set of expectations about the country that influenced my subsequent choices of subject matter for my art. Having grown up in the city, I sought a change from the sensory overload, the heavily industrialized and developed urban landscape that I had always known. I wanted, perhaps naively, to live near "the country." My expectations were romantic: that there would be more harmony with nature, more access to wilderness. My vision of rural America most closely resembled a giant state park. This desire to experience a new setting played a major role in my decision to do my graduate work in rural southern Indiana, and in my subsequent choices of motifs for my artwork.

Initially, I was attracted to the motif of cows, in part because of the shape relationships in their bodies, such as the balance of their bulky forms on narrow legs and the hippopotamus-like shapes of their heads. After years of studying the human figure, I was drawn to cows' unique anatomical characteristics and spent several months drawing them from life. The time I spent drawing cows acquainted me not only with their physical forms, but with aspects of their social interaction with each other as well. I realized that they were not the passive and calm stereotypical "Bossy." These dairy cows were far more aggressive than I had imagined, and I observed them frequently butting each other, and also mounting in imitation of sexual activity. Their cycle of artificially inseminated pregnancy and lactation was evinced in their swollen bellies and udders, and their mud-splattered genitals reminded me of William Butler Yeats's "Crazy Jane Talked to the Bishop": "love has pitched his mansion in the place of excrement" (line 15). Prepared to expect sweet and gentle cows, I was unprepared for the more challenging aspects of bovine behavior.

My preconceptions were based on images of cows from such sources as advertisements and picture books, most of which promoted a sanitized image. With no firsthand knowledge of farm animals, a compilation of animal images of human invention set me up for a rude awakening at the actual experience. The disparity between the more disturbing aspects of the actual animal and the pleasant aura surrounding the re-created animal became more obvious as friends gave me gifts that used cow imagery. On coffee mugs, cards, and notepads, patterns of the black and white spots of Holsteins had become a popular feature in surface design, suggesting a bucolic and unproblematic wholesomeness.

The contrast between the romanticization inherent in cute cow paraphernalia and my own experience with these gritty and aggressive beasts prompted my interest in life-sized displays of animals frequently seen on suburban lawns and in garden shops. As with the cows, the disparity between the real and the re-created extends to wild animals rendered in concrete, which present none of the challenges of coexisting with the real animals.

My experience from working in the landscape had primed me to look closely at how references to the wild or natural world are incorporated into the suburban landscape in the form of painted concrete statuary. New housing developments where land has been cleared of trees and brush are a frequent sight in the suburban landscape. The greenery that is replanted is carefully controlled in the form of manicured shrubs and clipped lawns. I find both irony and poignancy in these new developments where young trees from the nursery and concrete geese or mushrooms decorate newly sodded lawns. Living animals may have had their habitat destroyed by suburban development, and they are now re-created as concrete statues. Actual wildlife may be regarded as a nuisance or may even have been driven away, but their images live on, preserving the form without the content and its attendant complications.

The juxtaposition between the real and the reconstructed becomes painfully clear if we consider the difference between the various concrete beasts that are a fairly common sight on the suburban highway landscape and the real animals, struck by passing cars, that lie dead by the side of the road. These images speak of our capacity to romanticize that which our life-style actually threatens and destroys. A clear relationship exists between the challenge to see the destruction of animal life and habitat and the desire to re-create nature in a fantasy form. Symbolically, we do not have to be concerned with the often-unpleasant fates of real animals, as we can reconstruct their images in the form of painted concrete statues.

Lawn art also reflects the need to feel connected to nature, but it is a nature of our own creation and one that we can control, a nature that complements our world as a backdrop without inconveniencing us. These statues suggest a denial of a highly mechanized and depersonalized society and evoke nostalgia for a more pastoral era when we had more day-to-day contact with other species. The three-dimensional representations substitute for actual wildlife and speak to a sense of loss for a connection that has been severed. Their re-creation for decorative purposes communicates a need to feel connected to the vestiges of animals, but that need takes the form of superficial representation. In 1987, for example, Miles Kimball of Oshkosh, Wisconsin, advertised ceramic squirrels in its mail-order catalog. The copy read: "Cute and curious creatures that lend lots of winsome appeal to your backyard, and they promise never to raid the food in your bird feeder or wake you with their chattering."

The "Scampering Squirrels" came in three sizes and presented none of the problems of the real animal. Popular statuettes of deer also imply whimsical and pleasant associations with *Bambi*-like woodland scenes, and create an almost fairy-tale feeling. Ironically, these statuettes appear in areas where wildlife may encroach upon suburban developments, prompting homeowners to complain about deer trampling their gardens. I can attest to this firsthand, as I simultaneously was thrilled to glimpse a deer in my backyard and dismayed to see the leaves sheared off the tops of our pole beans.

Looking critically at lawn art confronts us with a view of animal life that is manufactured, remains under our control, and fulfills a need for connections with animals but with none of the unpredictability or risks presented by the animals themselves. In this arena, we may destroy and re-create nature at will. Although a particularly large, garish, or unusual display may arrest the attention of passing viewers, the geese in a flower border or deer or a frog on a lawn are intended to affect the setting around them, not to have an affect themselves. Unlike other aspects of landscaping such as bushes and flowers, concrete animals are of wholly human manufacture, purchased and displayed at some cost and effort of the homeowner. The fact that we accept these decorated statues as fixtures of the landscape that are nearly as ubiquitous as bushes or flowers on front lawns demonstrates how accustomed we have grown to proprietorship over nature.

My interest in such yard art and the prints I have made based on this subject may prompt us to consider cultural assumptions about nature and our place as humans within it. Other chapters in this book have posed a range of questions about the environment and humankind; I want to end my own chapter by letting my art pose some additional questions. Printmaking entails the making of a surface such as a metal plate or a woodblock from which an image may be printed in multiple copies. Prior to the invention of photography, it was a means for commercial reproduction of images as well as a medium for artistic expression. As a printmaker, however, I am less interested in the medium for its potential for multiple copies of a finished image than for the unique drawing vocabulary of the process, the mark making, and the tactile characteristics.

My chief areas of focus are etching and drypoint. In these techniques, the design is incised on a plate, usually of copper or zinc, by means of a sharp instrument or by a corrosive agent such as acid. The printed image is produced by rubbing ink into the lines and off the smooth surface of the plate; the image is then transferred to dampened paper by means of a high pressure press. The instruments and materials play an important role in the process and the appearance of the finished print. The process of corrosion, pitting, or scratching that produces the printed image is an integral part of the finished work. Because of the three-dimensional nature of the image surface, etching and drypoint produce an image that is tactile as well as graphic.

Lawn and garden outlets, and also the factories where the statuary are made, have offered me opportunities to observe a large selection of lawn objects isolated from their typical context. The animals and the icons of a more spiritual nature, such as Madonnas and cherubs, are arranged row after row for purchase; this arrangement intensifies the commodification of images. The ordered ranks of concrete animals in various sizes and shapes suggest a regimentation that further increases the distance between living animals and the forms of their concrete reproductions.

Juxtapositions of animals, religious figurines, and classical statuary as well as unreal scale relationships (a frog may be as large as a deer) suggested a dreamlike or surreal atmosphere in an early work of mine, de-

Figure 1. Ellen Price, *Little we see in Nature that is ours* (1988). Drypoint, 15 inches × 21 inches. Collection of the artist.

picted in figure 1. The scratchy, tentative, and almost evanescent quality of the line augments this surreal atmosphere and emphasizes how far removed these figures are from the actual world. This work is executed in the drypoint process in which a needle scratches the surface of the metal plate to produce a line that holds ink when printed. Areas of light tone are produced by the use of fine abrasives such as sandpaper or emery cloth.

I have titled this piece *Little we see in Nature that is ours*, from William Wordsworth's poem, "The World Is Too Much with Us." For me, the poem voices the loss caused as much by alienation from nature as by an obsession with commodities:

> The world is too much with us; late and soon,
> Getting and spending we lay waste our powers;
> Little we see in Nature that is ours,
> We have given our hearts away, a sordid boon! (Lines 1–4)

The poem, written in 1807, seems prescient of what may arguably be one of the most pressing issues of our age, that is, how we interact with our environment. We do, as Wordsworth says, see very little when we see animals as decorative appendages that may be purchased in any size or shape and placed where we want them.

I did several works in drypoint on collections of lawn ornaments similar to this piece. Those works share certain compositional aspects such as the tentative line quality, pale tonalities, and shallow and ambiguous space where the statuary are primarily related to each other rather than to a clear or obvious architectural or outdoor setting. By focusing on the bases to which the figures' legs are attached I have emphasized both their frozen stances and their origins. Humans, though not physically present in the images, are implicated as their manufacturers and consumers.

Later intaglio works drew on similar subject matter; however, statuary depicting religious icons played a more central role in later work, where figures of cherubs and Madonnas interacted with other statuary. For example, *Symbols of Faith*, an intaglio print (see fig. 2), utilizes the juxtaposition of two statuettes of the Christ figure with a collie, along with some other concrete beasts. Here, the placement of this statuary raises issues about the relationship between these display objects. For example, in the art of many nonindustrialized cultures, animal imagery is incorporated into the worship of deities; ritual and worship includes the acknowledgment of human dependence on other species. The collie sitting at attention suggests a faithful guardian, and the Christ figure represents a guardian as well as a savior. In this scene both religious and secular images represent belief symbols that serve a similar purpose.

In figure 3, an etching entitled *Concrete Beasts*, the forms of the statuary are massed in a compacted and shallow space with smaller-scale figures across the top of the picture plane providing the sole indication of depth. The deer, attached to a square base and perched in the birdbath, are rendered with a heavy thick line that silhouettes their forms. The use of dark tonalities establishes a brooding atmosphere; here the flattened creatures make up a strange zoo. The corrosive process of etching produces lines on the metal plate as well as the pitted, scratched surface texture of many of the background marks on this print. The textural quality produced by the pitted and scratched surface suggests our eroding relationship with other species.

The life-sized reproductions of animals that pose on streets and highways fulfill a cultural contract as substitutes for reality at the same time that they reflect a desire for ownership over nature. The acceptance of the

Figure 2. Ellen Price, *Symbols of Faith* (1989). Soft-ground etching, 32 inches × 23 inches. Collection of the artist.

fantasy of happy woodland and barnyard creatures persists in spite of common knowledge of the exploitation of animals and the destruction of their habitats. Concrete beasts are a physical reminder of the distance between the human and nonhuman worlds, and the reconstruction of their superficial form does not restore the wild, but rather manufactures an-

Figure 3. Ellen Price, *Concrete Beasts* (1989). Soft-ground etching, 24 inches ×
36 inches. Collection of the artist.

other commodity. The romanticized reconstruction of living animals denies the failings of our commodity-based culture with regard to the environment and the destruction of species. These concrete beasts may
reflect our desire for connection to nature, but it is a reconstructed nature
that presents no difficulties and mirrors none of our failings.

REFERENCES

Bresson, H. C. 1989. *Line by Line: The Drawings of Henri Cartier Bresson.* Preface by Julien
    Levy. New York: Thames and Hudson.

In the Beginning: Creation, Restoration, and Turner's *Genesis*

Judith de Luce

The Vergilian inspiration of modern scientific mythology is not hard to see.
. . . But now that politics and science at least are beginning to focus once
more on the moon, it is possible that a new construct will be formed, and
a new table of metaphors organize the imagery of our poets.

Northrop Frye, *Myth and Mythmaking*

The United Nations has sent "Chance" Van Riebeck and a team of scientists to survey Mars, but instead of following those orders, they begin
the process of terraforming the planet. This requires warming Mars, rendering its lethal atmosphere innocuous, and populating its barren surface
with living organisms.

At first glance, Chance and his team appear to be part of a project that
has recently found its way into the popular and scholarly presses and
which aims to make Mars habitable for humans (Broad 1991; McKay et
al. 1991). They are, in fact, actors in Frederick Turner's 1988 epic poem
Genesis, and the action described takes place over one hundred years from
now. Let me continue with the poem's story line.

On Earth, the reigning Ecotheists, led by Van Riebeck's estranged
wife Gaea, believe that humankind is as evil as its manipulation of nature.
Alarmed by what Chance is doing, the Ecotheists attempt to try the
renegades. The Martian colonists who survive the war that ensues
finally manage to obtain a record of all the genetic material of Earth and
with it continue the process of "gardening" Mars, inspired in part by
Beatrice, daughter of Chance and Gaea. Finally, Beatrice gives birth

to Hermione, the great prophet who sings the new world of Mars into being.

In an age when we encounter relatively few epic poems and even fewer cosmogonic epics, *Genesis* comes as a pleasant surprise. I am tempted to examine its ten thousand lines in depth; to consider the choice of genre and meter; to evaluate the elaborate interplay of allusions, from the navy hymn and Dante to Lévi-Strauss and Pompeiian frescoes; to examine the use of imagery, the development of character, and the dissonance created by unexpected shifts in tone.

Such a literary analysis, however, would detract from the focus of this book. In some ways a discussion of *Genesis* could appear redundant here; after all, Turner's essay "The Invented Landscape" in this volume is nearly a prose version of *Genesis*. In that chapter, Turner suggests that terraforming another planet is a logical, appropriate step for humans: "Inventionist ecology, to give a crude definition, maintains that it is both possible and desirable not only to conserve natural resources, preserve natural ecosystems, and restore natural landscapes, but also, when the occasion warrants and the knowledge is sufficient, to *create* new ecosystems, new landscapes, perhaps even new species" (Turner's emphasis). Turner goes on to say that "the . . . reason why we should not be too anxious about the idea of the created landscape is that we will not stay on this planet forever. We must conserve and preserve the life of the Earth, but there are also dead planets out there that might be brought to life without risk. NASA is already seriously researching the proposition." One needn't read *Genesis*, then, to learn about Ecotheism or to explore different varieties of "gardening"; we need only read Turner's lead essay.

The poem prompts questions about terraforming, however, with the immediacy and irresistible drama peculiar to myth and to poetry. It is one thing to read an explication of the facts and figures and techniques of terraforming; it is quite another to confront directly and to participate albeit vicariously in the process itself.

There is another reason to supplement the perspectives of scientist and historian with a discussion of this poem. As Jean-Pierre Vernant has observed, "Poetry faces, in one direction, the world of praxis or action, a world of events occurring in time. In the opposite direction, it faces the world of theoria, of images and ideas, the conceptual or visualizable world spread out in space, or mental space" (1980, 203). Moreover, poetry by its very nature encompasses a vision that transcends a particular instance. For this reason Aristotle argued in the *Poetics* that poetry was more "philosophical" than history: "a poet's object is not to tell what actually happened but what could and would happen either probably or inevitably. Poetry is something more scientific and serious than history, because poetry tends to give general truths while history gives particular facts" (ix.1-3). We may employ computer modeling to learn how to terraform Mars, but only poetry can confront us with the implications of such an act.

Finally, if we do succeed in terraforming Mars, we will need to manufacture myths and rituals to account for the process, to remind us of why we did it, and to confirm our relationship with what we created. We may well need this very *Genesis* as the narrative foundation for a series of rituals to underscore this latest act of creation.

Having said all of this, I propose to analyze certain features of *Genesis* as an opportunity to reflect on some questions about restoration. At the outset I must admit that I cannot do justice here to this remarkable poem. Rather, I have selected for comment those features of the poem that I find most provocative and which, taken together, encourage reflection.

Genesis has much in common with such creation myths as the Akkadian *Enuma Elish*, Hesiod's *Theogony*, and parts of the biblical Genesis, as well as the stories of the Yoruba, the Fang, and the Hopi. Within Western tradition, at least, definitions of myth, helped along by the change from oral to written literature (Vernant 1980, 186), came to be stated in terms of what myth was not: myth as not true, myth as not logical. For our purposes I suggest we reject dichotomies and define myth as a narrative created by a group to explain that which cannot be explained any other way. Such a narrative includes agents who are superhuman and does not participate in the time and space of history, but "exceeds the boundaries of history" (Otzen et al. 1980, 7).

Myths are neither lies nor simply curious stories. On the contrary, they often provide the truest and most serious accounts of itself that a culture can give. Reflecting our fears and our expectations, myths confront directly those questions that Gauguin asked with his 1897 canvas, *Where do we come from? What are we? Where are we going?*

In a sense, every myth is a creation myth, showing "how a reality came into existence, whether it be the total reality, the cosmos, or only a fragment—an island, a species of plant, a human institution" (Eliade 1959, 97). Cosmogonic stories, those stories that tell of the creation of the cosmos and the suppression of chaos, tend to enjoy a privileged position in a culture's mythology, perhaps because such stories deal with questions of ultimate importance. How was the cosmos formed? What are the implications of that creation for human existence? Are there limits to human action? Are we responsible for or to the cosmos?

The actual details of cosmogony vary substantially from one culture to another, but generally speaking, the creation of the cosmos entails the ordering of undifferentiated matter. This ordering may appear as a biological metaphor, as vegetable or human reproduction. The narrative may be as elaborate as an animal tale or as simple as a description of like elements clinging to like. There may or may not be a divine "creator" responsible for the ordering. Conflict among generations may well be incorporated into the story; epic battles may ensue, or sophisticated political stratagems may be required to complete the process of creating order out of confusion (Otzen et al. 1980, 16).

The act of creation does not end here. The very process of composing a cosmogonic tale, indeed every creative act, "repeats the pre-eminent cosmogonic act, the Creation of the world" (Eliade 1954, 18). Repetition of the story of that preeminent act re-creates the cosmos yet again, and such performance or repetition is often an important part of a culture's rituals. The ritual repetition of the *Enuma Elish*, for example, in which the Babylonian god Marduk defeated Tiamat (salt water) and formed the universe from her body, "repeated . . . actualized, the cosmogony, the passage from chaos to cosmos" (Eliade 1954, 56). The repetition of the story at New Year did more than refresh people's memories of the exploits of Marduk; it reasserted the act of creation and also the reality of the cosmos itself. The worldview of myth recognizes that existence is "determined by the tension between cosmos and chaos. In short, it was the task of cult to reinforce the cosmos and combat the destructive forces which assail it" (Otzen et al. 1980, 59). Both Turner and William Jordan speak of ritual performance in the same breath with proposals for restoring landscape. In mythic terms, we could argue, for example, that not only does ritual planting restore a particular prairie but the very act of planting restores the entire earth.

In spite of what Northrop Frye expected of poets as they responded to space exploration, Turner has not invented new metaphors to write about terraforming Mars so much as he has relied on and subsequently transformed the rich tradition of world cosmogonies. In so doing he has created not only earthly life on Mars but the narrative techniques to talk about that life. In fact, the redactor of *Genesis*, toward the end of the poem, reflects on the challenge Martian poets will face as they try to write about what has happened:

The poets of Mars must make the myths from scratch,
Invent the tunes, the jokes, the references;
They must be athletes of the dream, masters
Of the technology of inventive sleep,
Architects of the essential shades of mood.

(4.5.165-69)

Turner's creation poem encourages us to reflect on the prospect of ter-raforming and its attendant ethical issues in part because *Genesis* both fol-lows closely and departs significantly from traditional cosmogonies. That is, the tension between what we have come to expect in a cosmog-onic poem and what we actually find in Turner prompts us to take more seriously the possibility of accomplishing ourselves what Chance sets out to do. Hesiod's *Theogony* and the biblical Genesis represent two of the better known cosmogonic myths, and they inform Turner's poem most frequently and most compellingly. To underscore just how far Turner de-parts from tradition and to set the scene for his own creation story, I want to begin with these traditional cosmogonies.

The connection between Turner's *Genesis* and the eighth century B.C.E. *Theogony* of Hesiod goes beyond the obvious, such as the corre-spondence of the name Gaea/Gaia. After an expanded invocation, Hesiod recalls that ancient time when "First of all, the Void came into being, next broad-bosomed Earth [Gaia], the solid and eternal home of all" (II, 116). Hesiod recounts the ordering of the natural world and the hostilities that attended the appearance of each new generation of gods until Zeus estab-lished himself in the third generation.

After her initial appearance, Gaia produces Uranos (Sky), the moun-tains, and the seas by parthenogenesis. Then she sleeps with Uranos and conceives the Titans, including Cronus, Rhea, and Prometheus. But Uranos does not allow the children to see the light of day, so Gaia con-spires with her son Cronus, giving him a jagged sickle with which he castrates his father and assumes control.

Cronus and Rhea produce the second generation of the gods, including Zeus, Hera, and the other Olympians. Cronus is no better a father than Uranos, however; he swallows each child at its birth. After hiding the new-born Zeus, Rhea tricks Cronus with a stone wrapped in cloth. In time, Cro-nus vomits up the swallowed infant gods, and Zeus usurps power from his father, thereby becoming the "father of gods and humans" himself.

We can see the poem's focus shift gradually from female to male deity, from Gaia to Cronus to Zeus. There has been another shift as well. Conflict between fathers and sons in each succeeding generation would not have been unfamiliar to Hesiod's audience, but the violence of Uranos's and Cronus's overthrows does not repeat itself precisely in Zeus's generation. Rather, Zeus engages in rudimentary negotiation, not just brute force, to establish himself.

In the language of the King James Version of the Bible, "In the beginning God created . . . ": the first creation story in the biblical Genesis starts when God assumes the role of creator. According to this account, in orderly succession God creates and then judges each newly created feature of the world to be "good," from the light and the firmament to the herb–yielding seed and whales. On the sixth day, God creates humankind and grants this newest creature a special relationship with the creation and its creator. Humans receive "dominion over the fish of the sea, and over the fowl of the air, and over the cattle, and over all the earth" (Genesis 1:26).

In the second biblical account of creation, Adam appears before creation is completed: after the first man has appeared, God creates hosts of animals. Adam participates in the further ordering of the cosmos when "the Lord God . . . brought them [every created thing] unto Adam to see what he would call them: and whatsoever Adam called every living creature, that was the name thereof" (Genesis 2:19). That is, using that ability traditionally regarded as unique to humans, the ability to use language, Adam names the world that he sees around him. By naming it, he can understand it and therefore control it. Much of the significance of the biblical book of Genesis resides in this relationship between God and humankind, between a creator and a people chosen to enjoy a particular relationship with the intelligence responsible for ordering chaos into cosmos. Adam does not formally create anything himself, but his naming supports the process of ordering, and in that regard he is directly involved in one aspect of creation.

Turner begins his *Genesis* with a compelling "Listen!" Unlike the authors of Genesis and the *Theogony*, however, Turner does not recount a traditional tale created by a group: this is not, in the strictest sense, a myth. What is more, Turner does not tell his creation story as a straightforward narrative of a past event. Rather, the creation he describes will not take place for a century into the future, and the tale he "tells" is in fact the work of a poet who has yet to be born, "an underground poet living in New York about a hundred and twenty years after the events of the

poem under the gentle totalitarianism of the Earth's theocratic government" (from "Dramatis Personae," Turner 1988, 9). The redactor of
Genesis, then, transmits a story about the future that took place in the
past for the narrator of the poem. This complex chronology serves a useful purpose: "the poem may be designed as a warning to past ages of the
consequences of their fear of the future; on the other hand the action of
the poem may be a kind of performative invocation designed to bring
about the new choices it describes" (Turner 1988, 8).

Just as the conception of time and its relationship to creation has
shifted, so the identity of the creator has shifted, too. In Turner's *Genesis*,
humans have assumed the role of creator (I would say "arrogated" were
it not for the implications of that word) no less than we have assumed
that role in countless laboratories. This changes dramatically the relationship that humans may have with the cosmos. In the *Theogony* and in the
first creation story in Genesis, humans were uninvolved in the creation
process; they appeared well after the cosmos had been ordered/created. In
the second Genesis creation story, rather than appearing at the end of the
creative process, humans appeared in medias res and participated in the
subsequent creation. Whereas in other cosmogonic tales we learn what
relationships with the cosmos humans may expect, in *Genesis* humans
themselves have to determine what relationship they should or can have
with that which they have created.

In Hesiod, we saw the recurring hostility between fathers and sons,
and the shift from a female to a male deity. The Gaia hypothesis notwithstanding, surely one must recognize the Greek Gaia in Gaea Van Riebeck.
The Greek Gaia conspired with her son to battle with his father. The Ecotheist Gaia conspires with her son against his father/her husband and his
sister/her daughter. In Hesiod we saw the initial fertility of Gaia produce
much of the physical world but the generational strife between fathers
and sons determined who held power. In Turner we encounter several
generations as the story unfolds, from Chance and his daughter, taking
over Mars against the wishes of Gaea and other Ecotheists, to Beatrice,
directing the gardening of Mars, and the Sibyl, Hermione, singing the
new landscape into existence.

Finally, this is no creation ex nihilo or de novo. *Genesis* derives not
only from the tradition of cosmogonies but of apocalypse and eschatol-
ogy, as well; from the Götterdämmerung and various cataclysms that destroyed a flawed creation and set the scene for its restoration. The world
that Chance leaves behind as he begins the process of terraforming is one

in which humans have been poor stewards. He does not create afresh; rather, he restores, translating the environment of the Earth to a new site and repeating the course of evolution. Turner even connects the Martian restoration with the biblical flood and that subsequent restoration when he describes the spaceship Kalevala, the "tree" ship in which the Martian colonists will transport the Lima Codex back to Mars. This Codex contains a record of all of Earth's genetic material, without which it would not be possible to make Mars habitable for humans. In Genesis, Noah had loaded an ark with representatives of every living thing at the instruction of a god who established a covenant with him, promising that he would not flood the world again. In the Greek tradition, a man and a woman (Deucalion and Pyrrha) were responsible for repopulating the devastated earth after the flood. In *Genesis*, in the aftermath of a cataclysmic battle, a brother and sister locate and load the Lima Codex onto the ark for the restored world that will appear on Mars.

In *Genesis*, Turner implies that this restoration will be some kind of utopian creation, that humans will somehow avoid the problems that they created in their original home. Moreover, according to the author, Hermione "reconciles the ancient mystical wisdom of the Earth with the new science and cultural experience of Mars" (Turner 1988, 15). Yet at the same time that humans restore the landscape, the narrator of *Genesis*, writing after the terraforming had been completed, speaks of the degradation of the Earth in his own time and asks, "What was there left for my own ruined planet?" (5.5.279). If this means that humankind was unable to maintain tolerable life on Earth, but had to turn more and more to the terraformed Mars, then what does that suggest about how humankind will treat Mars?

A brief digression is in order here. If the landscape on Mars is somehow to be utopian, then one model for this new, restored landscape might be Greek Arcadia; Turner writes of Arcadia in "The Invented Landscape." In *Genesis*, in passages notable for their lyricism, Chance and then Beatrice derive inspiration from Arcadia, but I see a problem with this. I am reminded of my own first visit to Arcadia. I recall sitting with archaeologist friends in the numbing heat of a July afternoon, listening to the insects and the bells of unseen goats. In the stillness of that afternoon, we could well believe that the god Pan still walked the land. But this particular spot in the Peloponnese is one of the few places in Greece so isolated that it affords no glimpse of the sea. Arcadia is associated with the wild goat god, with anything but civilization, urban living, technology. It does not represent a solution to the conflicting demands of modern living, or any compromise between nature and culture or preservation and restoration. The absence of the sea reminds us how far Arcadia is removed from the commerce, the politics, the social realities of Greece, ancient or modern. The reality of Arcadia challenges us to consider whether we can afford a model of such profound isolation as we invent other landscapes.

I need to explore one more feature of *Genesis* before I can pose my series of questions prompted by this poem. The poem constantly echoes a variety of literary and artistic traditions. This constant touching base with the past underscores Turner's departure from that past and prepares us to ask questions of the future. I will limit my comments here to Turner's use of Vergil and Dante.

Once he had made it more or less safely from burning Troy to the shores of Italy, Vergil's Aeneas went to the Underworld to see his father, Anchises. Not only did Aeneas's trip take him through his own past, but Anchises showed him the future of Rome. Aeneas emerged from the Underworld arguably better equipped emotionally to continue on his mission to found Rome. Aeneas had as guide to the Underworld the great Cumaean Sibyl who was well known for the ambiguity of her oracles. The token for the trip (the passport, if you will) was the golden bough. Centuries later, Vergil would escort another traveler, Dante, through the *Inferno* until Beatrice took up the role of guide and teacher through the *Purgatorio* and the *Paradiso*.

In *Genesis*, Beatrice, daughter of Chance and Gaea, wields a golden bough, guiding the Martian colonists as they "garden" the planet. Once again Turner keeps us off balance by shifting the familiar details. The Trojan hero needed the golden bough to gain entry into the Underworld in defiance of its laws; in *Genesis*, a "gardener" of cosmic proportions uses the bough to lead the way into the mysteries of transforming/terraforming Mars. It would have been gratifying had the Sibyl been Beatrice; then the descent from Vergil to Dante to Turner would have been neat and convincing and simple. But Turner does not make his poem simple. Beatrice's gardening skills play a major role in the terraforming of Mars, but it is actually the Sibyl, descendant of Vergil's great seer, whose prophetic singing brings Mars to life much as the Hopi Spider Woman's song animates that world. Hermione reminds us that some things do not lend themselves readily to prose or to scientific discourse. Vernant was right: sometimes you need poetry. Technology is not enough; computer modeling is not enough; Mars is not truly habitable until the Sibyl sings her song. And just as Vergil's Sibyl helped guide Aeneas to the Underworld where his father could show him the future, so this Sibyl guides us as readers to consider the future possibilities on Mars. Would it be too extravagant to wonder whether the notorious ambiguity of the ancient Sibyl's prophecy applies here? Surely we must be very wise, and very cautious, if we are to understand fully the song of the Sibyl.

Turner's story of the struggle between the Ecotheists and the Martian colonists, the application of technology to a planet unsullied by contact with humankind, the appropriation of pristine planet by humankind . . . all of this poses some important questions. What if we are able to terraform Mars? What are the implications of rendering the environment of one's home planet so intolerable that moving to another planet is preferable? Are we facing the nonmythic equivalent to the flood?

Now that we can assume the role of the cosmic creator, how do we play our part? If we muff our lines, if we turn in a poor performance, as some would argue we have done on our current planet, then where will we go? Where will we turn our creative impulses and our technology?

Will we need *Genesis* as the creation myth of a Mars inhabited by humans? Will we need to repeat this poem annually, to remind ourselves of the motives for terraforming and of the conflicts that gave rise to the move from one planet to the next? Aeneas traveled from one end of the known world nearly to the other end to fulfill his destiny and found Rome, a feat that required enormous effort, as Vergil tells us. But *Genesis* has us leaving our home planet and traveling to yet another. Doesn't this *Genesis*, by virtue of its constant echoes of other traditions, remind us that we have heard this song before? In fact, we have heard this song in various languages, at various times in our existence as humans. The question remains: What have we learned? We have the technology, but have we the ethical basis for yet another creation, one that we will initiate? If we succeed in terraforming Mars, then surely we will need many myths, many rituals to explain not only how it was done, but why and at what cost.

REFERENCES

Aristotle. 1965. *The Poetics*. Trans. W. Hamilton Fyfe. Cambridge: Harvard University Press.

Broad, William J. 1991. "Can Mars Be Made Hospitable to Humans?" *New York Times* (October 2): B5-6.

Eliade, Mircea. 1954. *Cosmos and History: The Myth of the Eternal Return*. New York: Harper and Row.

————. 1959. *The Sacred and the Profane*. New York: Harcourt Brace and World.

Frye, Northrop. 1960. *Myth and Mythmaking*. Ed. Henry A. Murray. Boston: Beacon Press. 131.

Hesiod. 1953. *Theogony*. Trans. Norman O. Brown. Indianapolis: Bobbs-Merrill.

McKay, Christopher P., Owen B. Toon, and James Kasting. 1991. "Making Mars Habitable." *Nature* 352: 489-502.

Otzen, Benedikt, Hans Gottlieb, and Knud Jeppesen. 1980. *Myths in the Old Testament*. Trans. Frederick Cryer. London: SCM Press Ltd.

Turner, Frederick. 1988. *Genesis, an Epic Poem*. Dallas: Saybrook Publishing Company.

Vernant, Jean-Pierre. 1980. *Myth and Society*. Trans. Janet Lloyd. Atlantic Highlands, N.J.: Humanities Press.

The Little Hut on the Prairie: The Ritual Uses of Restoration

Ann Cline

> Ritual allows those who cannot will themselves out of the secular to perform
> the spiritual, as dancing allows the tongue-tied man a ceremony of love.
>
> Andre Dubus

Notwithstanding the avalanche of print to which this volume adds its
slight bit, the modern professional and academic often stands "tongue-
tied" before the spiritual dimensions of his or her world. The modern
hope has always been that with a few more studies, with the right
projects, with another experiment, we might stand in some kind of permanent favor with the destiny of nature, our own destiny assured. To
this end, ritual has seemed an irrelevant or frivolous part of our prepos-
itivist past.

William Jordan's proposed prairie restoration ritual and Frederick
Turner's observation that ecology has become a modern theism position
ritual in a new light of possible relevance. Neither Jordan nor Turner
spells out what a restoration ritual would entail, but their ideas are attractive: rituals can be a powerful way of discovery and are useful in propagating a worldview; rituals can involve experts and nonexperts alike in a
participatory way; rituals carry a power, or perhaps a mystique, absent
from most modern activity. There are risks to ritual making, however,
that may cause well-intentioned effort to slip into unexamined dogma
and sentimental frivolity. Jordan is right: tinkering with *awareness* is the
goal of ritual, ecological or other. In his view, if we let ordinary people in

on the process of restoring an ecological landscape, through a process he calls ritual, then many may come to share the awareness of the restoration scientist and will then, presumably, accept the tenets of the new creed proposed by Turner. However, if this creed remains unexamined and ignores significant reservations, such as the four Turner enumerates, its ritual can slip into blind dogma. In addition, if our guiding metaphors are merely emotional, we risk slipping into sentimentality or kitsch.

When Jordan refers to environmentalism's "ritual value, and to the crucial role of ritual in mediating" healthy relationships to nature (see " 'Sunflower Forest': Ecological Restoration as the Basis for a New Environmental Paradigm," this volume), we can picture a Grand March of the sort Milan Kundera (1984) describes: group sentiment aroused by the replay of key images that are no longer examined. Communist May Day celebrations and Miss America pageants, with all their forced cheer, are sentimental rituals awash in kitsch. In Jordan's case, the appeal to health could also arouse a preconditioned sentiment — one Turner would associate with "the unspoken principles of the ecological religion" (see "The Invented Landscape," this volume) — that is then connected to the condition of the environment and its human interventions. Though not necessarily harmful, especially when kitsch is recognized as such, to leave Jordan's call to ritual potentially close to kitsch is not as helpful as it might be. The harm that kitsch does, generally, is to exclude the views of skeptics, views that, in this case specifically, might expose the limitations of an unexamined ecotheism and might posit more effective alternatives than those espoused through mere sentimental stimulus/response. My purpose is not to challenge Jordan's and Turner's work or their intentions, but to reposition the idea of an ecological ritual in an unsentimental way, in a way that allows the skeptics — those who "cannot will themselves out of the secular" — an avenue to "perform the spiritual."

The potential risk of sentimentalization in a prairie ritual comes largely from the unexamined baggage of our positivist past and present: the notion, for example, that tinkering with the modern world will make it sustainable. This new creed holds as its primary principle of faith that humans *can* sustain themselves indefinitely on this planet: this is our mission, our destiny, the focus of our faith. For skeptics who cannot will themselves to faith, rituals of preservation and restoration might allow us all to perform the offices of the church. Before launching headlong into a program of ritual activity, however, it may be worthwhile to probe the assumptions at work, to allow the skeptic a slightly longer audience before his or her baptism. Suppose, for example, one posits a contradictory view of human destiny, one that fundamentally challenges the hopes of positivist and ecological faith. Suppose, for a moment, that human destiny is not to become caretaker or perpetual-care gardener for planet Earth, as if our world were some sort of giant, endowed, changeless cemetery. Suppose, instead, that the destiny of human beings is to do just the opposite; suppose our fate is to process our planet, much as termites process logs fallen in the forest. Our job, like theirs, is to render all that we are capable of consuming into a kind of compact powder, or dust, or sand. A person holding so unfashionable a view, sensing the glares and animosity of the audience, might well point in his or her defense to the seemingly "hard-wired" human trait of greed, the tendency of humans to extend consumption well beyond mere self-preservation into the elaborate and often dizzying heights of cultural and personal pleasures by which we tend to measure the quality of civilizations. If it were the goal of human destiny only to garden the earth in perpetuity, then why would we come so well equipped to imagine, invent, pursue, and enjoy the pleasures we are able to create beyond those required for sustenance?

The questions raised by the unfashionable skeptic may be contentious, but they are profitable. If, in our ritual of restoration, we cannot account for so basic a human quality as "greed" (to put it in the Anglo-Saxon and largely theological term, though the more psychological, French "pleasure" I will employ as well), then we have created a ritual of unexamined sentiment, rather than a ritual of substance that includes helpful skepticism. Let us not shun the skeptic too quickly; if we consider the skeptic's position a moment longer, the most significant corollary to human-as-termite can emerge. Even if the destiny of humans is to process planets, the skeptic would add that there is not the slightest need for us to hurry.

We probably have all the time in the world. If time is a human invention, then processing the planet unto the extinction of the last human sipping a last martini *is* All the Time in the World. The value of the skeptic here is to remind us of the forgotten quality of ritual, about which both the ecologically faithful and the skeptic can agree: rituals take a lot of time. This is one of the reasons why people in a hurry hate them so much. What might be done in a matter of moments—a chalice of wine is consumed by a priest after eating a bite of bread—is dragged out for hours. This is annoying to the truly greedy, but to the termites who realize that Time is all we have, the longer the better, the longer our consumptive destiny will continue. The irony of ritual is that by consuming more time than materials, Time itself—that is, global human existence—is extended. This is one way the skeptic might restate the "healthy relationship to nature" of which Jordan speaks.

Modern ritualists often make much of the symbolic content of ritual. Those who favor revitalization of such traditional rituals as those practiced by various Native American and African peoples often point out the contemporary and sometimes ecological significance of these rituals' symbolic content. From the point of view of time, however, the content of ritual is beside the point. If the point of ecological ritual is to prolong Time, especially the time spent consuming things, and if ritual is going to slacken the pace of the greedy, then it is more critical that ritual prolong pleasure than that it engage symbolic meaning. In fact, it could be argued that as the Protestant Reformers and later positivists purged ritual from what was becoming modern life, what they purged was ritual's symbolic meaning and the authority of those institutions invested in and insisting upon their own interpretation of meaning, a purge necessary for both groups. It is no wonder then that there are still plenty among us who hold an aversion to symbolic ritual, and would do so even for the benign intention of ecology.

The skeptic might fill out the ritual intimated by Jordan and Turner without a priestly caste or an imposing dogma. The skeptic's ritual might be more a dance in which the pleasure of one's being as a human, including those elements Turner describes, is engaged over "un-customary" time, that is, time prolonged beyond what would be normally required. In place of a heavy-handed, symbolic text based on the principles of the "Ecotheist" creed, one might imagine instead an event whose symbolic or actual "calls and refrains" echo Turner's four reservations about the creed itself. Or, one might imagine an event that positions us where Turner might have us be, and then explores the dimensions or the pleasures of that existence. In fact, such a ritual already exists, and before returning to the more specific idea of a prairie restoration ritual, it may be helpful to examine this precedent that does much of the work the skeptic would call useful.

The ritual I will describe begins in anticipation when the host sends invitations, and replies are returned. At the appointed time, the arriving guests proceed along a garden path to an arbor where they await the arrival of the host, who has meanwhile arranged the details of the event. When host and guests have greeted one another, the host returns to a small hut near the arbor. The guests follow one by one along a different path to a stopping point marked by an arrangement of rocks resembling a mountain spring. Each guest stoops to rinse his or her hands, then walks to the hut, and enters. Each goes first to a small alcove in which the host has provided a graphic message, the "call and refrain" of the ritual text. Then, if the ritual is to be preceded by a light meal, the guests seat themselves and await the arrival of the host, who brings in trays of several small dishes, one tray for each guest. Wine is then served, and if there are to be several courses, these are brought out one at a time. After the meal is concluded, the guests retire to the garden arbor where they may smoke tobacco. All the while, in a corner of the room, a charcoal fire heats water in an iron kettle.

In the guests' absence, the host changes the character of the room: the host may replace the text with flowers or change the quality of light by placing shades over the windows. The charcoal fire is given its final arrangement. The guests are then called back to the hut. (If no light meal is served, then the sequence from this point constitutes the entire ritual.) Upon entering, the guests inspect the alcove and then the fire and kettle. When they are all seated, the host brings the utensils into the room and continues the ritual process. In the first stage, several of the articles are wiped with a cloth and all are arranged in precise order. Next, a bowl is warmed with water taken from the kettle. After the bowl is emptied and dried, a portion of powder is taken from a container and placed into the bowl, where water is added and mixed. The bowl is given to a guest to drink, then returned to the host. A bowl is made for each guest in turn; the bowl is carefully cleaned between guests.

After the last guest has drunk, the utensils are cleaned and arranged for removal. The guests may then ask to inspect some of the utensils more closely. The host presents the utensils chosen for viewing, the others are removed, and then each guest in turn has an opportunity to examine them and question the host about their origin. Informal conversation may begin at this point and continue until the guests thank the host, depart, and retrace their steps through the garden.

What I have just described in a deliberately generic manner are the key elements of what the West calls the Japanese tea ceremony and what the Japanese call *Chanoyu* (boiling water for tea). The Japanese terminology is more helpful to us: this is an activity about process, the product of which is tea but the means of which focus upon the prior creation of boiling water, a process that cannot be hurried even by Western terms (a watched pot never boils). As we examine what happens during a tea ritual, we notice the salience of process and of pleasure. The use of a single bowl necessitates that one wait one's turn, a process of individual anticipation and remembrance as the event unfolds. Surrounding this are processes of deliberate laying out and care for utensils that are, within their cultural context, quite ordinary. In fact, the ritual of tea viewed through the eyes of its traditional creators, is a prolonged inhabitation of the ordinary. This occupancy of the mundane is not judgmental, in that there are few strict taboos; however, taste as evolved by the tea masters encouraged what could be called sensible or even inverted selection. Iron was used among the utensils and in teahouse construction only where no other material would do—a sensible selection. Precious gold, when it was used, might have been employed to mend a teabowl of ordinary ware but perhaps historic importance—an example of inverted selection. The point here is that the ritual was not dogmatic in instructing humans to be other than they were; it did, however, prolong a typical human event in a manner in which that state of being might be savored. In short, the Japanese tea ritual was a ritual of pleasure-over-time.

For our purposes, the variations in this format are not so important, except to note that they are vast, as are the differences in possible characteristics of the setting (Anderson 1991; Cline 1988; Sadler 1962). Even so, by tradition, the garden and the teahouse are spotless, refined, and "rustic," adhering to the Japanese aesthetic term *wabi*, or refined rusticity. The garden is neither colorful nor gorgeous, instead conveying the sense that one is walking along a mountain path. Rocks and plant materials resemble wild, native species; flowering trees and ornamental shrubs are avoided. The teahouse is small, frequently with a tiny entrance through which guests must crawl, and is made of the most ordinary building materials: in traditional Japan, wood, plaster, and thatch, although these are carefully crafted and maintained.

The reader might have guessed that I was describing the Japanese tea ritual, an event many people know enough about to consider quite foreign to the modern West. Yet I think that the elements as I described them are the basis for an event far less culture-bound than may appear when seen in their Japanese form. These same elements could form the basis for a ritual in which the "ordinariness" of one's circumstances may be played out, savoring a particularity in time and place. Even an ecological ritual might incorporate these essential elements. They could, in fact, form the basis for a prairie restoration ritual. One view of the Japanese ritual sees it as largely parasitic upon a deep cultural exploration of the life of the recluse and his hut—beginning with Lao Tzu's image of the reclusive individual who was "contented with his food, pleased with his clothing, satisfied with his home, taking pleasure in his rustic tasks" (Waley 1958, 241), later including the growing phenomenon of the recluse poet, which by the T'ang dynasty turned into a steady stream, and ending in the urban reclusion, or *shin yin*, of the Sung dynasty gentleman in his urban wilderness garden. Another view of the tea ritual finds a parallel genealogy running deep in the Western psyche as well. The philosopher Gaston Bachelard has written:

> The hut appears to be the tap-root of the function of inhabiting. It is the
> simplest of human plants, the one that needs no ramifications in order
> to exist. Indeed, it is so simple that it no longer belongs to our
> memories—which at times are too full of imagery—but to legend; it is a
> center of legend. When we are lost in darkness and see a distant
> glimmer of light, who does not dream of a thatched cottage, or, to go
> more deeply still into legend, of a hermit's hut? (31)

When the Japanese imported the recluse paintings and poetry of Sung dynasty China, typically depicting the recluse poet's world of a tiny hut set in a vast mountain terrain frequently crisscrossed by narrow paths along which one or two human figures amble or stand gazing at the view, they were able to place themselves and their own literary recluses, like Kamo no Chōmei and Yoshida Kenkō, or later their tea masters, such as Shukō, Rikyū, and Oribe, in the scene. This assured that the tea huts and their landscape gardens could be associated with prior, familiar inhabitants. If one would compare, side by side, the tea ritual's environment and a Sung dynasty mountain landscape ink painting, one would find striking similarities in both spirit and detail. It would be safe to say that the intention of the tea environment is to re-create the landscape of the mountain recluse poet, for the purpose of spending time in its basic expression of dwelling. Indeed, what sounds like a mere social occasion, and was so for the Chinese gentleman, became a precise and exacting ritual of landscape habitation for the Japanese military and merchant classes.

It therefore seems equally plausible to place in this landscape any number of figures familiar to the West: Grimm's "Fisherman and His Wife," Dorothy Wordsworth, Henry David Thoreau, Aldo Leopold, Harlan Hubbard, Gary Snyder, and now, William Jordan. From this position—the intellectual apart from the world, or what we might call "the termite in exile"—derives the necessary position for the ritual that interests us, one that evolves directly from the recluse's poetic impulse, from the position of deliberate exile to a landscape of "useless" cultivation. At this point in our journey, it is possible to draw a tentative though partial parallel between, on the one hand, the activities of the Chinese recluse poets and Japanese teamen and, on the other, prairie restoration and rituals, as suggested by Jordan and Turner. Even though Jordan is restoring a landscape known to have been present at one time on its site, and the teamen are creating new landscapes only loosely founded on prior ecological type, both the teamen and Jordan choose sites that are close to if not within the boundaries of urban life. In addition, both have created what does not necessarily strike the reader initially as a pretty garden. Both draw some advantage from proximity and plainness. In order for the rituals they are aiming for to occur, however, a dwelling is required: a "little hut on the prairie" in which the poet dwells and onto which the pleasures of the site are drawn first into human consciousness, then into poetic discourse, and finally into ritual.

The turn to ritual, then, relates the internal project of the tea master, to discover and employ sources of pleasure found in rather ordinary and mundane circumstances—Lao Tzu's idyllic state where one took *pleasure* in rustic tasks—to the termite's project of a ritual of pleasure-over-time in which the quality of consumable goods is parasitic upon cultural norms and their quantity is extended by being prolonged over time. In my own experiments with ecological ritual, I have built two structures for the contemporary recluse. In the first, a six-by-eight-foot hut, a freely translated tea ritual occurs; in the second, a three-by-fifty-foot structure, something quite different takes place. There I am attempting to surround the consumption of a McDonald's hamburger with several hours of activity, delay and pleasure turning fast food into slow food, and letting the ecological chips fall where they may. If the belief that the rain forest is being sacrificed for hamburgers is widely shared, what are the implications of sanctifying this sacrifice? Would this sacrifice be more tolerable if it proceeded at a snail's pace? Does this position us somewhere between evolution and where Jordan and Turner might have us be—not unaware of the implications of our actions, but circumspect about a creed that would deny our essential being and its role in change and evolution? The ecologically concerned could still argue that by acting out the pleasure of delay, the larger project of total consumption is affected positively. Yet the skeptic may also be assuaged: no priestly finger has pointed grimly to ecological disaster but instead our host and ritual have offered us a world of normal pleasure, enhanced by its prolongation over time. It seems to me that this is the work a prairie ritual ought to do. While the creation and maintenance of the restored prairie, or any other restored or created landscape, may be part of the environmentalist's project, the public may best join the landscape and the circumstances of its creation as guests, invited to dwell for a time in the landscape of delay and of pleasure. By paralleling ordinary life, this invitation has direct resonance with its guests and may therefore resist sentimentality and kitsch and instead approach significance.

To understand what this significance is, it may be useful in closing to return to the case provided by Japanese culture. Although it cannot be stated that the tea ritual was responsible for the successful achievement of the Tokugawa period, this two-hundred-year era should interest any ecologist because it was a time when a densely populated nation was stopped, by internal restraint, from further foreign trade or expansion and spent its time as a kind of hothouse of near stasis. One might call this time the Tokugawa Spaceship. If it were my purpose to discuss only the ecological implications of the Tokugawa period, we would have to acknowledge the political power struggle that successfully hindered the aspirations of the merchant class and various sumptuary regulations that accompanied this dampening (Shively 1964/65). My purpose, however, is to reflect upon the nature of the satisfactions that were achieved within the Tokugawa Spaceship, satisfactions that were largely the result of pleasure-over-time most thoroughly explored in the practice of the tea ritual. The effect of this ritual in Japan was first to explore the nature of time-released pleasure, then to encourage a vast territory in which this pleasure might be enjoyed, and finally to inspire a transformation of an entire life-style. Indeed, the function of this ritual in Japanese life appears to be its demonstration of pleasure, which could still be achieved despite limitations placed upon it, limitations that though partly a matter of status quo politics anticipated those that ecological circumstances would have required.

In our own situation, I would like to argue that a similar exploration of pleasure-over-time and its ritual landscape and habitation could begin to offer similar benefits, not the least of which is the very pleasure derived from the process. I would suggest that whenever there is to be a prairie restoration ritual, we refrain from speaking of its ecological value or function and emphasize instead its pleasure. That will be enough to keep us busy for several generations, and after that will come what comes after that.

REFERENCES

Anderson, Jennifer L. 1991. *An Introduction to Japanese Tea Ritual*. Albany: State University of New York Press.

Bachelard, Gaston. 1964. *The Poetics of Space*. Trans. Maria Jolas. Boston: Beacon Press.

Cline, Ann. 1988. "The Tea 'Ceremony' as a Contemporary Medium for Design Inquiry." *Journal of Kentucky Studies* 5 (September): 129-41.

Dubus, Andre. 1987. "A Father's Story." In John B. Breslin, ed., *The Substance of Things Hoped For: Short Fiction by Modern Catholic Authors*. Garden City, N.J.: Doubleday Books. 147-67.

Kundera, Milan. 1984. *The Unbearable Lightness of Being*. Trans. Michael Henry Heim. New York: Harper & Row. 247-53.

Sadler, A. L. 1962. *Cha-no-yu: The Japanese Tea Ceremony*. Rutland, Vt., and Tokyo: Charles E. Tuttle Company.

Shively, Donald. 1964/65. "Sumptuary Regulations and Status in Early Tokugawa Japan." *Harvard Journal of Asiatic Studies* 25: 123-64.

Waley, Arthur. 1958. *The Way and Its Power: A Study of the Tao te Ching and Its Place in Chinese Thought*. New York: Grove Press.

The Poetics and Politics of Prairie Restoration

Constance Pierce

What do restorations restore to us? Certainly not the past. And yet that is their enthusiastic promise: to restock the present with the physical and even biological evidence of the past, and to maintain convincing settings that invite us to reinhabit various phases of a largely unproblematic history.

That is to say that restorations, especially public restorations, are at once an enrichment and a theft. It's true that restorations give us something unusual that we would not otherwise have. But what is restored is never something that we've lost, so much as something we've imagined that we've lost, and imagined in a most contemporary way. Nothing is restored. Something is presented, with an eye toward history. Sometimes this is an intelligent and imaginative eye, and the result of its effort can be an instruction and a delight. But such a realized vision is (of course) never the lost thing of our yearnings. Nor could it be. Denying the laws of time is the purest folly, and almost nobody really participates in this doomed venture. We know that restorations are not "the thing itself," and we judge them on their approximating power, their attention to detail and fact, their abilities to transport.

Always and inevitably, the thing that is restored comes into being through the neglect or erasure of something else, and too often the thing lost is something very much in our interest to have and to know. The restorative arts, for all of their fretting about history, have traditionally been prettifying ones, especially if the product was for public consumption. Even when coerced or shamed into adding back what's been left out (as Williamsburg has begun to "validate" the lives of slaves and women in its version of the history of eighteenth-century Virginia), the unsavory is estheticized against even the best intentions by the very fact of its being put on display.

How could it all be otherwise? The restoration of anything necessitates selection and interpretation. These are the characteristic features of collections, the characteristic forms of memorial projects in general, the characteristics of history as a genre, even as revisionists busily cut new career paths through its thickets.

Restored prairies will be like this, too. Their specific evolutions from prairies into whatever they had become prior to the restoration will be plucked out as though not indigenous, which is not true. And then—in the Frederick Turner/William Jordan vision—these new-old prairies would become the sites of rituals. Some of these rituals, too, will have been recuperated unproblematically from the past. They may or may not be prudent behavior for the twenty-first century (hunting has been mentioned), but they will be new-old rituals. Given what Turner and Jordan have said in print and in person, no doubt these ceremonials will be promoted as "natural" and therefore unproblematically good.

Is my critique . . . churlish? Turner refers to restoration ecology as "noble," and because I eventually wish to evoke class considerations, but more consciously than he does, I might as well play churl to his noble from the outset.

At first, it is difficult even to see an important issue in prairie restoration. It seems like a harmless avocation. Who cares if several or even a hundred plots of otherwise unused ground are gardened in this unusual way? We've suddenly metamorphosed into a nation of gardeners anyway. Some of us even cultivate prairie grasses and meadows in our yards. Gardening, transforming our little bits of real estate into something that expresses our Edenic visions, or promises to relocate us away from our troublesome historical circumstances, is just one more sign of these very private times. Gardening, among all else it might be, is another expression of alienation and longing for dominion over *something*, even if it's just the backyard. We are private citizens practicing land reform, attempting small antidotes to so much remote control and the overwhelming sense, by now, of nonnegotiable power-at-a-distance. And so we move earth around, and we erect perennial borders, against the perception of that other, apparently perennial, border between us and some inchoate Them.

With so many alienated fast-trackers taking time to grow the roses, what's one more garden? And why not a restored prairie? We are in no danger of being overrun by restoration ecologists. Shouldn't we let them restore and advocate in peace?

But if nothing is at issue, how has what looks like a "debate" over restoration ecology got so well under way? Why the voluminous position papers? The publications of Jordan and Turner alone on restoration ecology and related matters would fill several books. Perhaps the issue is money. Some part of the issue must be public money.

Acts of restoration carried out in the name of the commonweal are a kind of robbing Peter to pay Paul. Usually, someone who is not-us (well, not a Martian, but someone who is not particularly representative of the public mass) determines that a place or thing is worthy of preserving or re-creating, and then goes about eliciting our cooperation. Often this is without our knowing it; it is our confounded philanthropic (i.e., taxpaying) selves who are cornered. Money is also enticed from our schoolchildren through patriotic or sentimental rhetoric, parting them from their nickels and dimes until these accumulate in a monstrous sum. Then one fine day something (often something monstrous, like a warship) is restored and opened for our inspection.

With restoration, as with much else in our national life, we literally buy into some nebulous someone's sense of the heroic and the ennobling, a "consensus" of authenticities that are realized in space. But these sites never feel very authentic when we visit them as reflective adults. In fact, how often we are disturbed—or aghast—when we see what we've unwittingly helped to bring about or allowed to be brought about in our names. The past is often horrible enough, but the pretty past pandered to us in a semiofficial way denies our capabilities for evaluating more inclusive data. In short, it condescends.

Both Turner and Jordan, like magical creatures with special optics, look backward with one eye and claim to look far ahead with the other. Jordan envisions a brave new world of volunteer restoration ecologists loose on a prairie that he hints will be a kind of unspecified marketplace, while Turner (who has frequently used the word "explore" in his discussions of restoration ecology) is fixed on Outer Space, circa Jules Verne— and he begs the question of *why* we might have to set up camp on Jupiter or Mars some day in the future. Could the "environmentalists"—about whom both men have expressed reservations—have a point after all? That one day this planet might be uninhabitable?

Looking backward, in the most suspect way implied by the phrase, is one thing the Turner/Jordan model of prairie restoration seems to be about—however much Jordan wishes to separate himself from the preservationists, however futuristic and modern Turner seems in his rejection of both "purists" and "moralists." Turner's enthusiasm for colonizing space and his faith in a biological basis of esthetics are, by turns, entrepreneurial, deterministic, and utopian. Jordan (whose circumstances are different, in that he is dependent on public money) has perhaps needed to articulate a public use-value for the prairie of his dreams. But his description of grasslands where gatherings on primal weekends will set fires and traumatize "foreign" species, evokes some of the most unappetizing and violent American traditions.

I am curious, however, about the relationship between Jordan's view of pumping meaning into yet another museum display and something like the (partially) agricultural goals of the Land Institute in Kansas (Jackson 1980). Jordan, in his public remarks, has cited farmers as villains of the prairie at least as often as he has castigated the preservationists and environmentalists. And of course it was farming that originally disturbed or destroyed many of the native prairies. But by including agriculture in its restoration plans, the Land Institute refers to farming, which has an equally important history on the great plains. This is something virtually absent from Jordan's ideal, save in the incidental connection implied in "weeding."

I think it is important for anyone undertaking a historical restoration to try to acknowledge as complete a history as possible. It wasn't a frivolity that erased the original prairies, but a need for food and the need to establish an economy of crops. We might want to take issue with the way agriculture has been conducted in this country, but is it necessary to erase that history to restore prairies?

The Land Institute's conception includes a pragmatic possibility, that of actually producing food. Among the prairie grasses, wheat and other grains would be allowed to grow and to be harvested. This is an idealistic venture, too, but one with a much more compelling rhetoric for restoring a prairie than Jordan's proposal. Despite the drawback of the high initial cost of such a species of agriculture, the long-term goal of the Land Institute is one that *seems* environmentally sound. (I confess to not knowing, at the level of science, if this is indeed the case.) That this restoration undertaking also promises a desirable product no less basic than food, no less universal in its interest for people than food, can't help but engage some of the public's favor, perhaps even long term. Furthermore, it is straightforward about the fact that it aims at a new prairie.

As for the legions of recreational weeders and burners that Jordan envisions, I can't imagine that they would materialize. Given the recreation options of Americans, it's more likely that prairie maintenance would have to rely on occasional, or even one-time, tourists. But should such volunteers come regularly in useful numbers, then it seems that Jordan's plan could go ahead without a great deal of public money. In any case, he might want to take himself out of competition with the environmentalists. Though I share his skepticism about some of the interests that get subsumed under this wholesome rubric, "the environment" at its life-sustaining level has many pressing needs.

Turner's view is more disturbing—to my mind, decadent. I think he must sense a decadence in his own argument; otherwise, why would he try to discredit his critics in advance as "moralists"? There is something decadent in his rhetoric for "play," something worse in his estheticism, his uncritical enthusiasm for the "Arcadian."

In discussing American prairies, Turner has repeatedly said that he wishes to place them within the Arcadian tradition. That is to say, an aristocratic tradition, and a European tradition. (Interestingly, Jordan has criticized Yellowstone National Park as an example of our preserving an "aristocratic" version of nature.) Although restorations—even in America, even at public expense—often do speak the desires of an elite, to imply that the *original* American prairies had anything much in common with the grand estate gardens of Europe seems exceedingly odd, and unnecessary. In his long poem *Genesis*, Turner mentions the gardens of our own Du Ponts, the chemical aristocracy, as part of his scheme to recontextualize prairies within an American arcadia. But if the prairies are valuable, why do they need to be "elevated"? If they are Americana (more connected to the native populations than to the European farmers who destroyed them), then evoking the Arcadian tradition does not make sense.

Perhaps we have found a telling point of contrast between the original prairies and prairie restoration. What a restored prairie would have in common with the grand gardens of the moneyed classes is elaborate human design and an enormous requirement for labor. Gardens are by definition labor-intensive, and they are the result of human esthetics. American prairies were not designed, but restored American prairies, meticulously weeded of intruders and seeded with carefully selected vegetation, will certainly share a high artificiality with the great gardens of the world. Consider that artificiality along with the elite roots of many restorations, and you see something quite different from the "primal" and the "natural" that Jordan and Turner emphasize overtly when they write and speak about restoration ecology.

Part of what Turner wishes to say with his references to garden history may be that the American prairies are *no less* valuable than the great gardens of European aristocrats and American robber barons and tycoons. That is a "noble" democratic sentiment that many of us could be brought to share, and be enlightened in the process. But I am particularly struck by a bit of conversation between Turner and some contributors to this volume. During this discussion, I focused on his several references to the gardens at Stourhead. I said that I thought Stourhead was an insidious point of reference, because making those gardens had involved damming up a public river to create a private lake, never mind what farmers and fishermen and others who depended on the river might have lost in the process. Turner's cavalier response was, "But oh, what a beautiful lake!"

This kind of unapologetic estheticism is exactly what troubles me in Turner's discourse. For a moralist like myself, no amount of esthetic pleasure on the part of the gentry, eighteenth or twentieth century, could justify the loss of livelihood and comfort for the "peasantry" downstream—though their situation has not been detailed in any of the several garden histories of Stourhead that I have read. The price exacted from the people downstream is one of those things erased or left out, but I think it is very much in our interest to know about it.

Having said all this, I feel mean as an anthropomorphized snake in the (native) grass. To be entirely candid, I like the idea of having some prairie restorations, or re-creations. They would be a nice balance to the Rose Garden at the White House, not to mention all those English gardens springing up in the suburbs. I think it might be good to restore the non-native plants that eventually habituated themselves to the prairie, and to have some specific information about where they came from and how they put down their roots here. That seems a very American kind of story, worth including in the history.

What I don't like, obviously, is how Turner and Jordan think about what they hope to do, though I've learned much more from listening to them talk and reading their writing than this piece (of admittedly quasi-devil's advocacy) would indicate. I think Jordan is right, in principle, to propose a complement to the cathedralized nature projects historically preferred by the National Park Service; and Turner's thought moves as aggressively and adventurously as a space shuttle, even though I'm alarmed at some of the places it "explores," and some of the places it ignores. But I don't believe that a prairie restored according to the specific plans these men put forth could fail to be troublesome in the long run.

The mania for the past that has fallen on us in the United States—particularly during the past self-indulgent decade, when private restoration (of houses, lofts, and so on) surely reached some unprecedented level—coincides interestingly with our lost ground as a country with a future. And of course "the future of the planet" has been a worry for some time. Unlike Turner, I don't find any comfort in the idea of a future out on some Forbidden Planet, nor do I feel any enthusiasm for creating another ritual on this planet that is almost certain to feel empty of conviction. I'd rather we would decide to try, against whatever odds, to husband this planet, to focus our long-range vision on practicable ways to do this, when so much of the power (political and otherwise) is indeed at a distance, fatiguing us in advance of the effort.

Jordan mentions and dismisses Bill McKibben, but I think that McKibben's extreme point of view aims to be provocative, and that it is an appropriate response to much that is going on in our world. McKibben offers what he surely intends to be an alarming, and galvanizing, take on where we are, in nature and in culture. His recent *Age of Missing Information* (1992), in which he details the diminishing effects of "information" on our minds and spirits, seems right on the mark. But Jordan is right that there is something too gloomy about McKibben's laborious discourse. Not that it isn't credible, but it is enervating rather than energizing. We have to resist, even against what we might feel in our bones to be true. To do anything else makes no sense at all. But how helpful are prairies and rituals to the resistance? Just as difficult a question as whether nature is at an end, and culture a confusing and relentless mess of distracting uninformation, is whether rituals of any kind can be reinvested with significance and genuine function.

Besides, Nature hasn't ended. It has become a convention, with all a convention's cultural persistence—and all of its narcotizing power. This is the downside of McKibben's point of view. How do we now reinvigorate nature for ourselves? How do we make it immediate, and crucial, again—when the notion of its "end" is so large and deadening that it almost puts us to sleep, like that other unreality, the Bomb?

Some of the questions about prairie restoration, though also difficult, are more tractable. What are we leaving out of the history of this land in making such a vigilantly weeded garden? Would we be happy or terrorized if we were to show up at a prairie restored with our tax money, and therefore our implied endorsement, to find hunters stalking animals and a burning ritual in progress? Whom do we diminish in order to restore a prairie, and for whom do we do it? Is it just, or even in our self-interest, to forgo funding a chemical cleanup or an inner-city gardening project that puts needed food on a poor person's table for a whole season, in order to create a leisure activity on a simulated prairie? What other environmental projects are in competition for the same money?

In other words, we need to learn, specifically, who's downstream from this endeavor.

REFERENCES

Jackson, Wes. 1980. *New Roots for Agriculture*. San Francisco: Friends of the Earth/The Land Institute.
McKibben, Bill. 1992. *The Age of Missing Information*. New York: Random House.

Gardening with J. Crew: The Political Economy of Restoration Ecology

Jack Temple Kirby

Checking his property from the seat of a three-wheeled sport cycle during a deep December freeze, a Butler County, Ohio, farmer spotted something that had escaped his notice before. In a thicket near a ravine, buried to its hubs in frozen mud and entangled with defoliated vines, was a derelict tractor—a McCormick-Deering "15-30" from the mid-1920s. Red-brown with rust, its lower extremities melded with gray clay, the contraption shone with beauty in the farmer's eyes.

At the next thaw, the farmer and his son drove their powerful 1985 Ford tractor (which cost more than $35,000) to the thicket, locked a chain to the McCormick-Deering's rear axle, and drew home to a spacious repair barn the decrepit treasure, which sixty years before had sold for perhaps $1,200. For the next three months, the farmer and his son spent most of their time on the McCormick-Deering: disassembling, searching for replacement parts, cleaning, painting. The barn became the rural neighborhood's male social center, as other men watched, offered advice, and contributed their hands and eyes to the labor. When spring farm work drew the men outside once more, the project dominated their thoughts and conversation. At last, about late May, the old tractor's engine responded to the rebuilt magneto. The "15-30" rolled out of the barn, valves aclutter, the beaming son at the wheel, the farmer and his neighbors an applauding audience. The job was completed in time for the great annual ritual of the larger community of restorationists—the Butler County Antique Machinery Club's show at Hueston Woods State Park, near Oxford.

At this show, each July, farmer-hobbyists from throughout south-western Ohio and neighboring eastern Indiana roll in with giant trailers burdened with Case steam tractors (ca. 1910), Fords, Fordson tractors of the World War I era, Farmalls, McCormick-Deerings, and John Deeres, plus long belts and power take-off attachments for threshing, sawing timber, even washing clothes. The farmers camp with their families in tents and trailers, socialize, show off their treasures, and tinker with them endlessly. The atmosphere evokes rural reunions of bygone days when, the crops laid by for the summer, relatives and neighbors camped, prayed, and frolicked together. Women centered traditional reunions, and they are the practical glue at Hueston Woods as well, but the imme-diate cause of this celebration is the masculine culture of mechanics.

In the Middle West this culture is older than the reciprocating engine. Here farmers have long known that laborers had the option of industrial work, that labor-saving machinery was a necessity, and that a successful farmer was also an ingenious mechanic. Nowadays, beleaguered family farmers seem more compelled than ever to improvise, so they conjoin their mechanical passion with practicality. The sixty-year-old McCor-mick-Deering may be more an objet d'art than an implement, but many forty- and fifty-year-old tractors are actually in service on farms, per-forming light cultivation or hauling, while large newer machines plow and spread chemicals. The men are proud of their accomplishments at restoration, much like aficionados of old automobiles. They choose to live their local history; and although a social or spiritual dimension to the work of restoration is not precisely articulated, both aspects are manifest in the repair barns and at Hueston Woods each year.

I am a regular at these old machinery festivals. I share the farmers' af-fection for the clicking of pre–World War II valves, and I am fascinated by the restorationists' absorption with the workings of gears no longer man-ufactured. The farmers and their handicrafts teach me, a historian of farming and rural life, what it was like in the recent past. Their annual show is the best museum I have ever seen. As a historian of the South-east, where the proprietary family farm has not been the dominant mode, and where mechanization is very recent, I gain perspective from the Mid-dle Westerners' self-conscious perpetuation of a deep regional tradition.

Yet the farmer-restorationists trouble me, too. They lavish so much skill, patience, time, and money celebrating what has grown monstrous, as the prices of the McCormick-Deering and the 1985 Ford tractors (even allowing for inflation) indicate. The monetary burden of such machinery, mounting dizzily since about 1950, has driven most American farmers out of the calling. The rise of petroleum-based agricultural chemicals—accelerators of the "Green Revolution," along with larger and more expensive machinery—has rendered dangerous indebtedness ruinous. The restorationists, seemingly without insight into this irony, go on worshiping the progenitors of their predicament. Nor do they seem to grasp the enormous environmental impacts of heavy machinery and chemically dependent monoculture. "Experts" from government and from machinery and chemical corporations devise for farmers strategies of moderation—not structural change—stiffening resistance to reconceptualization. Farmers quite accustomed to international complexities in, say, the determination of prices for their grain and hogbellies, will not think large about political economy, agronomy, or ecology. Their daily lives are profoundly local, their "history" merely a validating antiquarianism.

Next to these farmers flows a parallel stream of certain urban intellectuals, at least one graduate degree for each, with a propensity to think not exactly large but rather cosmically, and who utter a torrent of portentous words. They are the restoration ecologists. Both sorts of restorationists are aggressive manipulators of the earth, ready to employ machinery and chemicals in order to work their wills. Both are preoccupied with heritage and specific places. Both value community and community ritual. Both employ local history as social and spiritual binders. Both are essentially antiquarians. Neither thinks large about political, economic, or environmental matters. The parallelism—allowing for class differences in manner of expression and aspiration—is remarkable. Sad, too, I think, considering the intelligence and ardor of both kinds of restorationists.

In 1990 William Jordan and Frederick Turner brought restoration ecology to Miami University and Oxford, Ohio, with vividly memorable charm. Jordan, an ecological scientist at the University of Wisconsin Arboretum, had been practicing restoration for years, but had only begun to consider its broader meaning when he read Turner's 1985 essay "Cultivating the American Garden" in *Harper's*. The poet Turner proposed an "arrogant and manipulative" approach to human-altered landscapes, in contrast to the "timid" stance of most environmentalists, and Jordan the practitioner was stunned and emboldened. Why not employ the technology of manipulation to *restore* landscapes to their original state? Bulldoze away civilization's effects, then slay incorrect botanical intruders with herbicides. Once a small restoration is achieved, humans' role is to cherish and maintain a living outdoor museum. This may entail many volunteers weeding meticulously with linoleum knives, as with the restored prairie plots around Chicago. Such focused drudgery (what Turner, then Jordan, terms intensive "gardening") will reintroduce a ritual relationship between humanity and nature comparable to that of pre-Columbian Amerindians. A way will open to deeper human understanding of cosmic unity.

Fred Turner in person represents even better than his arresting poetry and prose his status of guru to restoration ecologists. His imagination soars—sometimes without discernible guidance—and inspires. Turner articulates aesthetic and religious sensibilities that confer higher meaning to the work of such scientists as Jordan. (The farmer-hobbyists lack a guru, but one must remember that they neither subscribe to *Harper's* nor care about communication of the "intellectual" mode; see Jellison 1993; Kirby 1987.) Turner would hone keen humanity's "aesthetic capacities." Silent on broad questions of economic system, he infers acceptance of a hegemonic commodity culture that regards nature collectively as natural resource. To Turner, humanity's best recourse is intensive gardening in small places, the restoration of what he and Jordan call "classical nature" and "archaic landscapes." Turner insists that "we can't really go back. We can't go back to nature." What we *can* do is restore our souls while cultivating little prairies and oak savannas. Such are, in one of Turner's more remarkable expressions, "the biological foundations of aesthetics."

In North America, "classical" and "archaic" landscapes are those that prevailed between the last ice age and the appearance of Europeans. European imperialism was profoundly biological. In addition to the pathogens that exterminated native peoples, the Europeans introduced—purposefully and inadvertently—horses, cattle, swine, honeybees, birds, and crop plants and "weeds," creating what Alfred Crosby (1986) calls a "neo-Europe" from the Arctic Circle to the Rio Grande. ("Neo-Europes" also appeared in southern South America, South Africa, Australia, and New Zealand.) Jordan and Turner, both neo-Europeans themselves, disapprove. In their own odd way, they resemble the great American succession ecologist, Frederic Clements, who had, in the words of his Oxford University colleague A.G. Tansley, "a prejudice against civilization" (Worster 1985, 239). The restorationists are rather more particular: they despise European *plants*, which have been in situ in America for only four hundred years.

This prejudice was most apparent in 1990 when Turner, then Jordan, responded to an excellent question posed by the Miami University plant ecologist John Vankat. "I'd like to hear you reflect," said Vankat, "upon both your emotional and your intellectual response to the presence of a successful exotic plant in a restored prairie." Turner first responded that when "exotic" plants appear, then "probably it's time that the prairie were burned." This is, of course, nature's own manner of maintaining something approaching monoculture in a prairie. Fire destroys "weeds" while regenerating prairie grasses. When Vankat persisted (what would they do, say, in summer, when burning is impractical), Turner averred: "We act for restored ecosystems something like the immune system of a body, keeping out foreign bodies and foreign objects." Jordan continued: "If you see sweet clover in the middle of green prairie, we want to get it out of there, and that's both an intellectual and an emotional commitment, because we started out with a particular agenda. A peculiar one, if you will, to bring the system back into the condition it was at some sort of arbitrarily determined time in the past. It happens to interest us." Some response, no more logical than winter scenes in farmers' repair barns. Restorationists do restoration because they like it.

I have already suggested what lies behind the eccentric joy of hobby-ism. The restoration ecologists ultimately aim to do nothing less than save neo-Europeans' souls. Salvation is accomplished by ritual, which re-connects humans with nature. Weeding newly restored summer prairies with linoleum knives is reverential, but rather lonely. Autumnal community prairie burnings are more important and churchlike. A local burn master signals the day for the burn, when moisture and wind conditions are appropriate for a complete community turnout (meteorological uncertainty precludes scheduling in advance). Businesses and schools close, as though it were a "snow day." The scene at a burn is, says Turner, "liturgical," a "festival where people would show up and . . . celebrate and bring picnics . . . there would be performances, shows, and there would be that wonderful sort of smoke, and so on." The experience "rewires . . . human evolution." The smoke of a burning prairie becomes "incense in church, if you will. It's been a very, very successful little symbolic thing. And it has not come off as silly at all."

I myself have volunteered to help burn the little restored prairie at Miami University's Ecology Research Center. The experience appeals. I like fire. It is dangerous and warm, destructive and regenerative. Perhaps fire connects me to my Euro-American ancestors in the South (nearly four centuries of them) who burned the woods for their own good reasons, conflating their agricultural, hunting, revenge-making, and aesthetic traditions with Amerindian practice. Nonetheless I seriously doubt if weeding or fire-tending will save my soul. I feel no need for such redemption, in fact, for I belong here, along with dandelions, bees, English sparrows, and countless other European invaders of long, long ago.

My ancestors' biological imperialism is not only a great fact, it is an *old* fact. How long, one must ask the restorationists, does it take before legitimacy is achieved? St. John's wort (the Algonquians called it "Englishman's Foot"), plantains, horses, hogs—and I—are considerably older in North America than, say, the current royal line is in Britain. I respect blue stem and other native prairie grasses, and honor (not romanticize) Amerindians. I do not celebrate (or condemn) Cristofero Colombo (or Colón, or Columbus). If there is a judgmental god, s/he has already taken care of the sailor from Genoa. I have no "racial" pride. (Pride is prudently condemned in holy texts.) Racism, I think, is sinful and stupid, too—even if expressed oddly, as when one color-group assuages misguided guilt by deifying another color-group (see Heinegg 1978; Cronon 1986).

Another great and old fact is this: large populations (in relation to land area and its carrying capacity) with commodity cultures tend to manipulate and degrade ecosystems. The greater the degree of technological and market "sophistication," the greater the ecological degradation. Small subsistence cultures tend to change and degrade ecosystems minimally. Faith and ideology shape behavior and environmental impact somewhat, but population and the existence of commodity markets seem to be the main determinants in environmental history. Europeans and Euro-Americans might find solace in a bit of perspective: once upon a time *our* ancestors, too, were part of small subsistence cultures, made minimal environmental impacts, and lived with appropriate awe of Mother Earth.

So what is to be done, at this late date? Jordan and Turner prescribe merely the enlargement of their quasi-religious solution, a mass movement of "young and old, black and white" gardeners-in-the-prairies. In Jordan's slides of prairie restoration work along Chicago's metropolitan freeways, however, I noticed only upper-middle-class white folks of middling age: the L.L.Bean/J.Crew crowd (to which I, too, may be said to belong). Whether or not restoration ever becomes a significant crusade, I suggest that it not be taken more seriously than the rescue of derelict farming machinery. The results of both sorts of restoration are charming and instructive, but insubstantial and distracting. Restoration ecology is an expensive self-indulgence for the upper classes, a New Age substitute for psychiatry. It distracts intelligent and persuasive people from systemic initiatives.

One hopes that most Euro-American intellectuals have made or will soon make peace with history. (The Columbian quincentennial is a propitious moment for those who have not.) Then they may focus upon ongoing demographic, economic, and technological developments that degrade the planet. These include population pressures in many parts of the world, especially Africa. In so-called developed countries consumer-commodity culture runs amok, demanding more and more electrical power, inviting more and more nuclear waste, creating more air pollution from fossil-fuel consumption, and accumulating Himalayan piles of undisposable garbage. In North America, how does a free, individualistic society based on the automobile and personal convenience cope responsibly with the problem of great space? What of our class system, wherein the poor live in aesthetically unappealing housing? Are the aesthetics of urban life of no concern to environmentalists? Furthermore, the poor live most in danger of the toxicity our culture produces. Bean boots and Crew coats and hats would not protect them from, say, southern Louisiana's deadly groundwater and petrochemical plant emissions. The restorationists, like Marie Antoinette, would have them eat cake—which is the equivalent of cosmic gardening.

The best one might say of restoration ecology is that, like the wrongheaded romanticization of Amerindians, it raises consciousness about our culture's contempt for nature. Consciousness thus raised, however, leads nowhere except to museums—or strange outdoor churches. Intellectuals (including scientists and poets) have a greater responsibility.

REFERENCES

Cronon, William. 1983. *Changes in the Land: Indians, Colonists, and the Ecology of New England*. New York: Hill & Wang.

Crosby, Alfred W. 1986. *Ecological Imperialism: The Biological Expansion of Europe, 900-1900*. New York: Cambridge University Press.

Heinegg, Peter. 1978. "Lessons from the Indians: Ecological Piety." *North American Review* 163: 66-69.

Jellison, Katherine. 1993 (forthcoming). *Entitled to Power: Farm Women and Technology, 1913-1963*. Chapel Hill: University of North Carolina Press.

Kirby, Jack Temple. 1987. *Rural Worlds Lost: The American South, 1920-1960*. Baton Rouge: Louisiana State University Press.

Worster, Donald. 1985. *Nature's Economy: A History of Ecological Ideas*. New York: Cambridge University Press.

PART V

RESPONSES

Sunflower Seeds

William R. Jordan III

In my essay " 'Sunflower Forest': Ecological Restoration as the Basis for a New Environmental Paradigm," at the beginning of this book, I outlined what I had the temerity to call a new paradigm for the relationship between human cultures and the rest of nature. The essential elements of that paradigm are:

- That human beings, like all other species, are inextricably linked with ecosystems everywhere, and that they interact with and influence these systems even when they are very remote.
- That no ecosystem can be fully protected from this influence, and that for this reason the conservation of any ecosystem in its natural or historic condition will ultimately depend on a deliberate program of ecological restoration to compensate for novel influences on it. Thus the best paradigm for the natural ecosystem of the future is not a protected area, like Yellowstone National Park, but a consciously restored ecosystem, like Curtis Prairie at the University of Wisconsin-Madison Arboretum.
- That the ritual experience is essential to negotiating all problematic relationships, including the relationship between culture and nature, and that one reason for today's so-called environmental problems is the failure of modern cultures to provide an adequate repertory of rituals for this purpose.

- That the task of restoration provides the basis for a system of rituals—actually a new liturgy—for exploring, defining, and celebrating the terms of the relationship between the human and the larger biotic communities.

That is at least the essence of the set of ideas I developed in my essay. Now, some two hundred pages and fifteen essays later, where do we stand? What, if anything, remains of the original structure?

It is interesting to note that the ecologists who contributed to this volume responded to my first point (my restatement of the ecological axiom of interrelatedness) in quite different ways. Some, like Dwight Baldwin, John Wierwille, and Kimberly Medley, were reasonably comfortable not only with this, but also with my insistence on making it the basis for a philosophy of relationship between humans and the rest of nature. Medley, in particular, in her account of restoration efforts in the Tana River National Primate Reserve in Kenya, draws from that experience an endorsement of my emphasis on the value of restoration as a way of learning about ecosystems, as well as support for my larger conclusion that this experience serves to enhance the relationship between local residents and the landscape in a way that is likely to prove beneficial to that landscape in the long run.

These more positive responses came for the most part from ecologists with some direct experience with restoration, or at least with a restored landscape. Perhaps understandably, those working with ecosystems threatened with destruction and concerned primarily with protecting them from further damage were less sympathetic. David Gorchov, for example, though he recognizes the value of partly restored Amazonian rain forests and acknowledges their ecological superiority over monocultural tree farms, prefers to emphasize their inferiority to the natural ecosystems they are intended to replace and concludes that preservation (rather than restoration) "is the essential core strategy" for conservation of such areas.

Orie Loucks takes a similar position, from which he proceeds with admirable forthrightness to the logical conclusion. Loucks naturally accepts the axiom that we (like other species) interact with ecosystems everywhere, whether we know it or not. But he prefers for practical purposes to make a distinction, admittedly wholly arbitrary, between what he calls "maintenance restoration" in mildly disturbed areas (such as remote forests), and out-and-out restoration in more severely disturbed areas.

Loucks's purpose in doing this, it would seem, is to ensure what might be called the intellectual preservation of nature. In other words, what he is concerned about preserving is not any particular kind of landscape or community of species, but rather an *idea* of nature as that which is not human and not influenced by human activities. This is a motive I can easily understand, and which I do not dismiss, even if I do not share it. I do think, however, that it is important to consider exactly what it implies both for the natural landscape and for our relationship with it. Both are evident in Loucks's discussion. He idealizes a landscape like the Ding-hu Shan reserve not because it has any particular ecological features, such as particular species or communities, but simply because—or precisely to the extent that—it has not been actively managed.

What this means is that the system has been left to drift ecologically, and as a result it has been invaded by exotic species. Loucks notes that about one-fifth of the plant species now present in the preserve are exotics, introduced mainly as a result of "accidents of centuries of casual human activity," but also as a result of "recent programs of 'restoration'" on surrounding lands. In other words, what most ecologists would oppose under other circumstances (the casual introduction of species into new habitats) Loucks celebrates at Ding-hu Shan, apparently because it is casual and unintentional; what he objects to is the introduction of species by restorationists, presumably because that was deliberate. (We must assume here that he is using the term "restoration" loosely, as restoration by definition eschews introduction of exotic species except in special instances.)

In the process, one must assume that native species have been lost— that is what ecologists are always telling us is likely when they are warning us of the Pandoran dangers of species introductions. But this apparently does not concern Loucks. "What does it matter if a pollinator has been lost?" he asks. What he means, of course, is that the reserve is large enough to allow him to suppose that if the species has been lost in one area, it is flourishing in another, somewhere out of sight. Still, it's an odd question to be coming from an ecologist who is obviously concerned about the conservation of species and natural ecosystems.

One consequence of the distinction Loucks makes between mild and severe damage is the unacknowledged changes in ecosystems, inevitably accompanied by the extirpation—and eventual extinction—of component species. In the end it seems to me the story of Ding-hu Shan is one of the best endorsements I have seen of my contention that Curtis Prairie, and not the wilderness preserve, provides the best paradigm for the conservation of ecosystems—actual ecosystems, not some version of our idea of nature—in the long run.

This is what Loucks's distinction implies for conservation: basically, the preservation of an idea of nature at the expense of actual species and ecosystems. His insistence on the distinction between mild and severe human influences on ecosystems has important implications in other areas as well. It has implications, certainly, for our relationship with nature in the most general sense; and it also has important political implications. Rightly viewed, human influence on the landscape is not a bad thing; it is, in hard ecological fact, the very basis and warrant for our citizenship in the land community. Loucks would take this away from us, or, perhaps more accurately, from *most* of us, because the inevitable result of this distinction will be that most people will remain blissfully unaware of human influences deemed to be subtle, leaving knowledge and awareness of such influences to a group of experts. I frankly think that a more egalitarian approach to the fashioning of the relationship between culture and nature is not only preferable but also more likely to be effective in the long run.

To move on to the second pair of ideas from my paradigm, the two related to restoration as ritual, I was surprised that several of the authors were prepared to dismiss this idea out of hand, without even acknowledging the large body of knowledge regarding ritual and the role it plays in human experience. Of course, people are free to do this if they like, even if it means ignoring a whole dimension of experience, one that is at the core of most cultural traditions. At the same time it is fair to ask what the implications of this dismissal might be, both for the natural landscape and for our relationship with it.

One premise of the thinking I outlined in my essay is that ritual is the way humans (as well as some other animals) deal with deeply problematic relationships, such as the relationship between the individual and society, between the community and its gods, or between the human community and the rest of nature. To the extent this is true, slamming the door on ritual and the ritual experience actually precludes a positive relationship between culture and nature. I see no reason for doing this, and I find in these essays no compelling argument for doing so.

At the same time it seems to me that there are very good reasons for *not* doing so. Jack Temple Kirby, for example, is impatient with the idea of ritual because he feels that "it distracts intelligent and persuasive people from systemic initiatives." The question, however, is just what those initiatives might be, what the roots of environmental problems are, and which ones have the most urgent claim on our attention. Kirby writes about environmental problems such as overpopulation and pollution as if they were essentially technical problems, but this is surely not altogether the case. One of my basic assumptions has been that problems like these are rooted in human ideas, values, and beliefs. To ignore this, and to attempt to change habits and patterns of behavior without addressing the inner, subjective basis for them, is likely to entail exactly the kind of top-down imposition of power that Kirby and Constance Pierce are so quick to condemn when it is applied to the control of sweet clover in a restored prairie.

The essential question here is how to change widely shared ideas, values, and beliefs without imposing change from the outside or from the top down. Ritual is at least arguably one way to do this. So, of course, is education, but education itself is a performing art and inseparable from ritual. In any case, what reason is there for rejecting ritual, with its fuller use of the alternative languages of music, poetry, visual display, and expressive action? Do those who dismiss the idea of ritual really suppose that we are going to find our way back into nature without these powerful technologies of communication and communion?

There are, of course, many real questions about the efficacy, the uses, and the risks of the ritual experience. These remain for me open questions, but they are also serious questions that are entertained by scholars and practitioners in a variety of disciplines. Certainly they cannot responsibly be dismissed by slighting references to Marie Antoinette or to ritual as a merely eccentric pleasure.

Fortunately, not all the contributors dealt with the issue of ritual by dismissing it. For example, I found Ann Cline's thoughtful linking of restoration-as-ritual with the classic Japanese tea ceremony to be both helpful and provocative. One question it raises, which I think will repay further investigation, is the relationship between the tea ceremony, based on an act of consumption, and the act of restoration, which is not consumptive but constructive.

A few other points, perhaps of misunderstanding, call for some response and clarification. Kirby finds both me and Frederick Turner to be haters of civilization because we espouse the weeding and management of some ecosystems in order to maintain them in their natural or historic condition. But of course neither Turner nor I believe that restoration should be carried out everywhere, to the exclusion of houses, factories, and libraries; it should occur only in some places, to ensure the survival of species and their integration into human life, which I dare say all of us regard as an amenity of civilization.

Kirby also characterizes restoration as "lonely" work, but I think the experience of restorationists suggests exactly the opposite. Restoration, like other forms of gardening, or, for that matter, like the rehabilitation of an old tractor, can be carried out alone, but it also tends to develop into a social activity. Indeed, this is already happening in urban areas such as Chicago, New York, and the San Francisco Bay Area, where volunteers by the thousands have begun to come together to carry out high-quality restoration work on an environmentally significant scale. These events are rapidly becoming real community festivals—rituals, if you will—and should put to rest the notion that ritual of this kind is in any way an eccentric or peculiar taste.

To the observation that the people in my photographs of the Chicago restorationists were mostly from the "L. L. Bean crowd," I would respond that this seems a peculiarly limited data set from which to be drawing conclusions and passing judgment. In fact, restoration is turning out to be the basis for a distinctively egalitarian kind of environmentalism. People of all colors, and with a wide variety of shopping habits, are involved, and in fact restoration is providing the basis for intercultural environmental efforts in inner cities and on Native American lands.

Kirby, along with Pierce, suggests that restorationists are resisting change and should come to terms with history. I say that they are neither resisting change nor denying it, but rather studying it—and that in fact restoration, regarded as a performance, is among other things a way of exploring history and reflecting on its sometimes tragic and irreversible consequences. Certainly it is not an attempt to deny failure; rather, it is in part a kind of meditation on failure.

Carl Pletsch and G. Stanley Kane construed restoration as the assumption of sovereignty or dominion over nature. To me it looks more like the opposite: accepting responsibility for our influence over nature. Kane points out that, so far as the heuristic value of restoration is concerned, there are other kinds of knowledge than maker's knowledge. I quite agree, and would suggest that lover's knowledge, for example, is exemplified by the work of the restorationist, with its commitment to perceive the subject clearly in order to reproduce it faithfully.

Several authors suggested that I have tended to dismiss preservation as a valuable conservation strategy. I am sorry to have conveyed that impression. In fact, I see protection from novel influences as an indispensable first step toward the conservation of classic ecosystems—but only as a first step. To succeed, I think this must be combined with a program of, yes, continual restoration that acknowledges the inevitability of influence and consciously attempts to reverse it. My argument is not at all that restoration offers a substitute or even an alternative to preservation, but only that the two together are needed to establish a sound basis for the relationship between nature and culture. Further, I would argue that preservation and protection alone do not provide a basis for a coherent philosophy of nature and its relationship with human culture, but that the act of restoration (implicitly including a commitment to preservation) does.

In his essay, Loucks asserts that restored ecosystems are intrinsically less valuable—less interesting and spiritually less marvelous—than their natural counterparts. But this is merely an assertion of value, a personal conviction that he makes no effort to substantiate. An attempt to do so would surely take into account the immense heuristic value of the process of restoration on the one hand, and what is known about the ontology and etiology of spiritual value on the other. It would also have to consider the possibility that the value of an ecosystem is actually *increased* through our deliberate, constructive participation in its ecology, and that by transcribing natural systems into human understanding and concern, this process confers on them a kind of immortality. This, indeed, is closely akin to the process of making-sacred as described by Mircea Eliade and other historians of religion.

When Kane argues that the knowledge gained through restoration drives out mystery, I want to say that it does not drive it out, it only widens its frontier. (Surely willful ignorance is no substitute for mystery.) When Gene Willeke says restorationists' claims would be more credible if they included the less-appealing elements of nature in their work, I say they do. Restorationists may be credited with the aesthetic *discovery* of ecosystems such as the tall-grass prairie, which was widely regarded as ugly and desolate when the first restoration efforts began on the prairies a half-century ago, and with the rediscovery of fire, that unsentimental fact of ecological life that Pierce finds so repellent, so evocative of the Holocaust.

Finally, when Pierce and Kirby, pointing to the weeding and the fire, insist on the shamefulness of the kind of discrimination it implies, I say, yes—but all choices carry an element of shame, not least the decision to do nothing at all. The solution, as Turner has pointed out, is not to deny the shame, but to do something about it. And that brings us back to the subject of ritual.

The Invented Landscape (Reprise)

Frederick Turner

This volume has become an exemplary case study in the effects of paradigm change in the academy. The experience for one of the "target authors" is uniquely flattering, a little like the episode in *The Adventures of Tom Sawyer*, if I recall it correctly, when Tom achieves the widely held fantasy of witnessing his own funeral and hearing what everyone has to say. As a revenant, then, it is incumbent on one not to be too much of an anticlimax, while avoiding actions that might bring the townsfolk out with a stake to do the job properly this time.

What I primarily feel, however, is the need to express my appreciation. My thanks are due first to the editors. Carl Pletsch's profound understanding of Nietzsche partly underlies this book. He has seen the transvaluation of values that is going on in our ideas about nature, and brilliantly reinterpreted it in terms of the old debate over sovereignty. It is ironic and all too human that we should now so violently reject that sovereignty over the rest of nature on this planet, which for so long we yearningly but falsely imagined was ours by right—and at the very moment in history when that responsibility has been roughly and prematurely thrust upon us.

I would also like to thank Judith de Luce for her sensitive and generous reading of the poem *Genesis*. She has exactly caught the combination of Hesiodic tradition with revision of the tradition that the poet was aiming for (I say "the poet" because I do not regard the poem as "my own" work—I feel more like its redactor than its author). She has gone beyond close reading to a kind of "restorative" interpretation that leaves the

251

poem richer than it was before; this is exemplary, on the literary level, of the kind of relationship we can have with the rest of nature. My only quibble is that the Arcadia the poet had in mind is not that tumbled country of the Taygetus that we both remember from our sojourns in Greece, but the successive transformations or rereadings of that landscape by the pastoral tradition, both literary and horticultural.

Dwight Baldwin's essay, like his general contribution to the book as a whole and to the work of landscape restoration, exemplifies the mixture of humility and sensitive activism that is the sign of a truly constructive life. His hands are, in the best sense, soiled; his arms have been up to the elbows in the earth. He is aware of the complexities of that idiom; the world is always already soiled and fallen from what it was, but its very life and existence consist in the continued negotiation between return and radical transformation. When Wallace Stevens asks "Is there no change of death in paradise?" Baldwin knows the tragic but also profoundly hopeful answer.

In somewhat the same vein I would like to thank John Wierwille and Kimberly Medley, who both see ways in which the human presence in nature can be a constructive and enriching one; the differences in philosophical approach that may be found among those of us who share this conviction are energizing and positive.

Perhaps the most delightful surprise for me in this collection is Ann Cline's luminous and profound essay on the tea ceremony. This essay by itself would justify the daringly interdisciplinary approach that the editors have taken with this volume. I would like to take up her idea about delay and slowing things down at the end of this response, as it may, of all the things in this book, be among the ideas that deserve further research.

These writers—together with William Jordan, who in his own quiet way is a great visionary—appear to me to exemplify the fecund energies that can be released by a new intellectual paradigm, one that breaks the old categories, that switches figure and ground, that breaks frames and renders stock responses obsolete. But what makes this book especially interesting is that it also contains a classic suite of negative reactions to a new idea. New paradigms are not, of course, always right; but knowledge can only advance if such paradigms are understood and argued out in their own terms.

When attacking a paradigm that views all human activity as part of the natural evolutionary process, for instance, it does not help to simply continue the traditional critique of the human impact upon nature. It also rather misses the point to continue to argue, as Orie Loucks does, that there exist ecological "originals" of which humanly mediated reproductions can only be copies; part of the new paradigm is precisely the reminder that nature is already in the business of reproduction and copying, and thus the linear and dualistic distinction between authentic original and artificial reproduction is profoundly questionable. Far more germane to the defense of the old paradigm would be a plausible argument that there was some fundamental discontinuity in human evolution, some external unnatural influence that supervened upon it, so as to justify the distinction between the human and the natural, the artificial and the genuine. Those who accept the legitimacy of scientific knowledge would probably not accept the contention of the creationists that God created us separately from nature and planted us here on earth, but it is incumbent upon them to produce a substitute.

More helpful are those arguments from the point of view of the old paradigm that question whether, in lumping human beings together with the rest of nature, we are left with sufficient grounds to criticize the destructive activities of human beings. But those who make this argument must then deal with the new paradigm's own ways of making value choices. Ad hominem arguments about the questionable motives of the advocates of the new idea, or their possible conflict of interest, or their social class or gender or race or choice of clothing, do not contribute to the process of finding the truth and deciding what should or should not be done.

The new idea developed in this book is one that paradoxically clothes itself in old ideas, or rather, is a way of seeing old ideas in a new way. Innovation should not, perhaps, creep up on one so deceptively; it is so much easier to recognize when it comes with the familiar trumpet blasts of political dissidence, avant-garde style, and fashionable pessimism. Someone once said of the academy that its first response to a new thought is to say that it is ridiculous nonsense; its second is to say that it is ingenious but wrong; and its third is to say that of course it was known all along. What makes the new idea in this book confusing for those who believe in a linear view of history is that part of the new idea consists in a recognition that some of our old ideas—of landscape gardening and traditional land use—that have indeed been "known all along" may be part of the solution for the future.

It is fascinating to see how, when the old categories are violated, a sort of cognitive dissonance or category confusion sets in. Jordan and I are attacked on wildly different grounds: because we are too traditionalistic, and because we are devotees of futuristic technological progress; because we wish to restore the past, and because we refuse to acknowledge the inevitable horrors of the future; because we are too Eurocentric and because we would keep European weeds out of restored prairies; because we are dreamers, and because we are hard-eyed pragmatists; because we are irresponsible and because we accept too much responsibility. All these critiques may be valid, but perhaps a position that inspires such contradictory responses deserves a second look.

Both Jordan and I are attacked for being against ecological preservation; on the contrary, we are explicitly in favor of it. Our heresy is to add qualifications to that article of faith. For us, preservation remains a priority, but on the grounds of historical responsibility and professional commitment rather than metaphysical taboo, and for us restoration is one of the chief means of preservation. We see preservation in an absolute sense as an impossibility, because all things change by their own inner processes (succession, for example, in plant ecosystems). Thus one's legitimate desire for authenticity in nature must be satisfied in a different way from that of the preservationist Puritan. We find authenticity by selectively encouraging nature's own paradigmatic process of reproduction, using humanity's own natural aesthetics together with feedback from the ecosystem as a guide—a process that then becomes the real meaning of preservation. But neither Jordan nor I would cut down an old growth forest, nor would we condone such an action. Rather, we want to make it worthwhile for people not to cut down old growth forests, and thus make right action coincide with desire.

What perhaps confuses some of our critics is that our desire not to be punitive seems to take the rod of guilt out of the hands of the environmental Brahmins, while at the same time our recommendation that we all accept our responsibility, and thus the shame of our past actions with respect to the rest of the natural world, uncomfortably empowers the L. L. Bean-wearing, chablis-drinking middle classes that once lay so passively under the lash. More confusing still, perhaps, restoration on a large scale, which seems to be part of the Clinton/Gore proposal for national service, may be a way to give a creative role to the unemployed underclass, thereby depriving a considerable number of the brie and chablis set simultaneously of a clientele and a weapon of blackmail. Restoration is labor-intensive and deeply enjoyable; if it caught on we might have fewer unemployed to patronize.

Despite our disagreements, however, I would like to thank those writers who criticized my essay for their attention and for the many places where they clarified the lines of future argument and research. I would also like to comment on a few specific points.

Gene Willeke and David Gorchov, and others by implication, were skeptical about the capacities of human aesthetics as a guide to environmental practice. As working scientists, they are perhaps unfamiliar with the sophistication of the artistic and literary tradition, and think of ecological aesthetics as consisting of prettifying nature or reducing it to the look of a golf course. If this were what I meant by aesthetics they would be right. But there is no reason why restoration or creative ecopoesis should not take into its rich brew of ingredients all the best knowledge of molecular biology, soil science, evolutionary theory, ecology, and climatology, as well as human cultural history and local political sentiment, just as Leonardo included in his *Last Supper* a wealth of knowledge from philosophy, geometry, biblical history and interpretation, theology, and even contemporary psychological theory. The work of the artist Christo, whose remarkable interventions in the landscape involve all aspects of the local community, helps to show the way. Rational and empirical science has had its radical failures, and it should not, perhaps, be so arrogantly self-sufficient as to refuse the cooperation of the humanistic tradition in making decisions that affect the whole future of the planet. The values explored by the humanistic and artistic tradition—which is now increasingly a worldwide pancultural one—are not, as some scientists think, just a matter of individual taste, any more than is the choice of what scientific theory one accepts. The establishment of fundamental scientific principles is as slow and uncertain and revisionist a process as the establishment of an artistic canon, and those who live in glass houses should not throw stones. Oddly enough, when Willeke describes the work of his friend the river restorationist, and gently critiques his priorities, he is doing exactly what I advocate, that is, a fairly sophisticated aesthetic criticism.

Much more alarming is Constance Pierce's dismissal of aesthetics. Here it is a matter of professional competence; a professor of the humanities should know better than to reduce the whole world tradition of aesthetics to the merely pretty. For her aesthetics are dictated by economic and political power. Because she cannot believe that anyone could be so naive as to devote a lifetime to a kind of deep philosophical beauty that includes the tentative truths of science, she can only fall back on the insinuation that somehow Jordan and I have been bribed to say what we say. On the contrary: I am quite aware, and I believe Jordan is also, that if we said the politically correct things and did not rock the ideological boat we would both be considerably better off.

Pierce founds the only substantive part of her angry and sarcastic piece on an oral exchange whose subtext she unfortunately missed. She took exception to my description of the garden of Stourhead in Wiltshire, on the grounds that people living downstream of the artificial lake would suffer as a result of the whims of the gardeners. The context was important: we were surrounded by ecological scientists and other knowledgeable people who would recognize that a dam in the hilly and well-watered Wiltshire landscape would be largely a blessing for the downstream inhabitants, both because of its use for flood and debris control, and also because fishing at the outlet would be greatly improved after the lake had filled. She seemed to be under the impression that the lake would permanently interrupt the flow of water in the stream below. Not wishing to shame her by demonstrating her ignorance, I lightly passed over the issue, remarking that it was still a beautiful lake. For me beauty in its very essence would include such social factors as the benefit of other people. But for those who, like Pierce, believe only in political power, any other consideration, of truth, beauty, or ethical goodness, must be dismissed as epistemic control, escapism, mystification, or hypocrisy.

The problem with the idea of political power is that it is a profoundly linear conception. In definition, it is that the future state of one's sphere of action should include only what was in one's intention when one acted; and that one's sphere of action be enlarged without limit. In other words, one has political power only to the extent that one's actions are the unconditioned cause of the subsequent state of affairs. The problem is that one-way cause-and-effect relationships of this sort must always pay a thermodynamic penalty; the succeeding state of affairs will always be lesser in richness, complexity, and power (the capacity to do work) than the state that caused it. Thus political power always subjects itself to the entropic "running down" that affects all linear processes in the universe; indeed, we have seen what seemed like huge and impregnable accumulations of political power, such as the Soviet Union, evaporate after a few decades. Because the Soviet political structure could not open itself up either to nonlinear feedback within its own information system, or to nonlinear economic interchange with systems outside it, it disappeared

the moment its original ideological and physical resources had been exhausted. If one is to be creative at all, one must give up the urge to control, the paranoia about being controlled by others, and the unwillingness to undergo the consequences of one's actions.

One of the really interesting things that has been happening throughout a whole range of scientific disciplines, especially the ecological sciences, has been the emergence of nonlinear analysis of the relations between events and between the participants in a system—the body of
investigation and practice popularly known as chaos theory. Chaos theory deals with interactive systems that spontaneously generate their own
flexible, evolving, and beautiful forms of order, and confirms much of
the wisdom of the ancient classical aesthetics of all human cultures. In the
social sphere the new nonlinear approach offers a subtle and profound
analysis of collective human life—what Eastern European thinkers like
Vaclav Havel call "civil society"—that in my opinion largely invalidates
the simplistic and paranoid equations between knowledge and political
power offered by the likes of Althusser, Foucault, and Stalin.

Thus political power would seem to be a frail and flimsy idea upon
which to base the humanities; its explanatory effectiveness has failed in
both the natural and social sciences, which have moved on to subtler conceptions. It is a pity that academicians in the humanities like Pierce—and
also Jack Kirby, who, true Neo-Marxist that he is, yearns for the power
to reduce the population of Africa and block the world's peasantries from
further economic progress—are now adopting the linear conceptions of
power discarded as naive by other areas of discourse. But all is not lost. I
would direct Pierce's and Kirby's attention to recent developments in the
humanities that reflect the new movement of thought, for instance, the
work of Katherine Hayles, Elizabeth Fox-Genovese, and Alexander Ar-
gyros, and the new periodical *Common Knowledge*.

Stanley Kane makes several interesting points, some of which I would
like to answer briefly. He regards my position as arrogating to humanity
a role of domination, and argues with some persuasiveness that the fact
that we can kill our fellow species on the planet does not give us an ethical mandate to rule over them. One can agree with this philosophical
point, but the historian Carl Pletsch has, I believe, a wiser understanding
of our predicament. Our one-sided capacity to destroy without immediate reprisals against ourselves, and our historical exercise of that capacity,
do make us unique, and give us a sovereignty that, however little we
might desire or deserve it, and however unethical its source, cannot be

escaped. We can mitigate its effects by ruling ourselves; but we will only be good citizens of the planet and the universe if we do not sentimentally pretend to a state of equality that is simply not the case. As Ann Cline so trenchantly puts it, Kane's egalitarian vision of a harmonious "community" of human beings and herrings and mosquitoes and slime molds and herpes viruses is, alas, philosophical kitsch. "Greed," or the human desire for pleasure, has put us in charge. What Jordan and I propose are ways in which we can see ourselves so interwoven with the rest of nature, and our pleasures so bound up with ecological health and genetic diversity, that we will cease to be destroyers of the biosphere but instead will become its leading edge, its representative, its most conscious and aware and appreciative and activist member. We can do this only by interacting with other species, but this is exactly our point. When we so interact, we must recognize the inescapable advantages with which natural evolution has provided us.

Kane also argues against our contention that "maker's knowledge" is the deepest kind. I have already shown, in my response to Pierce, how "maker's knowledge" is incompatible with "controller's knowledge" or "domination"; this refutes Kane's argument that in urging an artist's attitude toward the biosphere we are seeking to dominate it. His counterexamples to the value of maker's knowledge (lover's knowledge, disciple's knowledge, parent's knowledge, enemy's knowledge, worshiper's knowledge, partner's knowledge) look plausible until they are examined. But a moment's reflection will show that lovers imaginatively create and re-create each other, that a disciple participates in the self-realization (and self-transcendence) of the guru, that a parent has a not inconsiderable role in forming a child, that we "make enemies," that the worshiper creates the god, that a partnership is made and does not burst miraculously into existence. Likewise, his examples of maker's knowledge would not be recognized as such by me. The sensitive and eloquent nervous system of the tortured victim is much more the "maker" of the torment than the torturer is; the parents and their genes are the makers of the baby, rather than the in vitro fertility expert. And the computer was indeed made by all those social users and analysts Kane mentions, as well as by the chip designer. Certainly the only way of truly creating (unless we believe in the power-universe of Constance Pierce) is by respecting and enlarging the autonomy of the world in which we are creating and the entities we create, and by seeking the widest collaboration in our creative effort; but that autonomy and collaboration are ours to give, as they have been ours to take away. Let us give them, then.

Ann Cline's vision of the tea ceremony seems to me to be exactly in the direction we need to go. What is especially deep about her idea is the role of delay. She understands the thermodynamic debt that the whole universe owes, and the fact that all life gets its organization, vitality, and beauty by speeding up the increase of entropy elsewhere. All dissipative organisms, termites as well as people, are parasites upon energy gradients and live by reducing them. Thus it is in our interest, and that of life on our planet, to delay the payment of the debt, to linger out and linger out our pleasures, to refine them by understatement, to enrich the inner fabric of time.

I would only add that I believe we can, in the breathing space offered by our ritual delaying tactics, "reinvest" what we borrow from the residual free energy of the Big Bang, in media that are not subject to the laws of thermodynamics—in other words, in the realm of (crudely) information or (more subtly) spirit. That investment, that bond or bonding, may draw a higher rate of interest, in all senses, than what we pay in terms of entropy. Thus we may generate time extensions that are, so to speak, at right angles to the time line of thermodynamics. Ilya Prigogine's thought points in this direction also, and it is one of the most exciting promises of the new paradigm. One bridge to that other time line is, I believe, our creative and healing work in the biosphere.

Conclusion: Constructing a New Ecological Paradigm

A. Dwight Baldwin, Jr., Judith de Luce, and Carl Pletsch

Large and innovative ideas are often known first by the opposition they provoke rather than the enthusiasm they generate. This may be true of ecological restoration and invention. William R. Jordan III and Frederick Turner argue that merely preserving what is left untouched of nature will neither rescue nature nor secure a future for humanity on earth. The philosophy (or theology) of preservation, in their view, actually inhibits us by preventing us from responding effectively to the ecological crisis we now face. They contend that we must assume responsibility for nature. At the very least we should be restoring portions of the earth that we have altered, and learning to think differently about our relationship to nature—beyond preservation.

The projects of Turner and Jordan overlap, but they are not identical. Jordan limits himself to restoring earthly landscapes; Turner considers constructing synthetic landscapes and even terraforming other planets. The two authors do concur in advocating an active, interventionist approach to nature, rather than a passive withdrawal from nature. They agree that we must acknowledge not only the impact our species has upon nature, but our membership in nature. According to them, we actually already are beyond preservation, but we have not yet accepted our responsibility to shape nature in positive ways.

In their essays here as well as in other publications, Jordan and Turner assert views that discomfort many environmentalists and ecologically minded people. Many of those who comment on their ideas in this volume—artists, biological and physical scientists, humanists, and social scientists—have reservations about the idea of landscape restoration, not to mention landscape invention. The views compiled here are not merely symptoms of the academic habit of testing new ideas by detailed criticism, however. They stem from deeply held and sometimes profound commitments to other views of nature and the human place in nature. Opposition to landscape restoration and invention is based on paradigms even older than the preservationist paradigm of conserving nature that Dora Lodwick traces to the late nineteenth century. For example, the conceptual dichotomy of nature/culture that separates humans from nature is as old as Western civilization. Turner and Jordan challenge the validity of such ancient and deeply ingrained habits of thought; no wonder this assembly of essays displays such a wide variety of criticisms of their views. If these criticisms are any indication, the idea of ecological construction is a very provocative one indeed.

The conflict presented here between preservation on the one hand and ecological restoration or construction on the other is between theories and even theologies of nature and the human place in nature. In the terms favored by Turner and Jordan, it is a conflict over paradigms, not practical projects. Neither Jordan nor Turner wishes to trammel or abandon the islands of untouched nature carefully preserved so far, and not even the sharpest critics of ecological construction deny the necessity of attempting to restore nature insofar as we are able. It is perhaps precisely because this is a conflict over paradigms that such a variety of questions must be raised about the ideas advanced by Jordan and Turner—from the scientific and technical to the ethical and aesthetic.

Scientists Gary Barrett, David Gorchov, and Orie Loucks all point out practical difficulties with the restorationist project. Whether these be the expense or technical problems associated with maintaining restorations like the Zurich bog noted by Loucks, the difficulties of reviving and reassembling even the most strategically cut tropical forest areas delineated by Gorchov, or the profound institutional supports that will be necessary before restoration can become part of an integrated ecological science as indicated by Barrett, these objections may be understood as challenges that must be met, not as arguments against attempting to restore landscapes and ecological communities. These scientists are not opposed to practical projects of restoration. What they do resist is the idea that the project of learning to restore and construct nature must take precedence over preservation. Loucks will not give up the idea that even approximately preserved nature—protected areas such as the forest at Ding-hu

Shan—is infinitely more valuable as "original" than any restored copies of nature. When Gorchov asserts that preservation is the core of the environmental project and Jordan responds that restoration is the rest of the apple, there is obviously an agreement to disagree about ecological paradigms. The critics acknowledge that restoration must be employed alongside preservation and other strategies of conservation. But they do not see the need to establish restoration as a new ecological paradigm, and they often vehemently oppose placing human intervention ahead of preserving pristine remnants of nature.

Others who reject ecological construction as a new paradigm include Gene Willeke, who acknowledges that vast tracts of abused land will require restoration but insists that this may be done without the philosophical framework proposed by Turner and Jordan. Willeke, Jack Kirby, and Constance Pierce all find ecological restoration a valuable complement to other methods of conservation but object to the semireligious aspects of the Turner/Jordan paradigm. In fact, the call for ecological rituals seems to have been either rejected or ignored by the majority of the contributors to this volume. To be sure, Ann Cline finds that ritual has played an important role in mediating human relations with nature, and that it may still have power to extend the pleasure of our interaction with the rest of nature. Judith de Luce relates the mythological narrative of Turner's poem *Genesis* to other creation myths, suggesting that if we do succeed in terraforming Mars, then we will need myths to underwrite our rituals and to remind us of why we did it and at what cost. Traditionally, the ritual repetition of creation myths has allowed humans to acknowledge the continuity of the universe, to assure themselves that the earth would remain fertile, animals would remain abundant for food and clothing, and so on. Ellen Price alludes to this also, in her discussion of the mundane rituals of yard work and the symbolism of lawn art. Curiously, neither Jordan nor Turner could see in any chapter except Cline's a reference to ritual, although their responses at the end of the book indicate that this is very important to them. Nothing could illustrate more clearly that this is a conflict over worldviews than these divergent attitudes toward ritual, myth, and religious language in the restorationist discourse. But the failure of the most trenchant critics to seriously engage the issues of ritual and myth does raise other interesting questions.

Is ritual objectionable here merely because we are out of the habit of thinking with ritual? Or because we live in a secular society accustomed to solving its common problems with the tools of science, economics, and politics . . . without reference to the religious? Or is the sort of ritual suggested by Jordan and Turner repugnant to some because it conflicts with other values of our civilization? To take another tack, is it possible that the economic, political, and technological rituals of our society are so ingrained in us that we are unaware that we already live by these secular rituals? Or that we actually have secular myths to underpin our society? Of course, this book does not attempt to compare the myths and rituals of our technological society dedicated to material progress, complete with its areas preserved from development, with the myths and rituals suggested by Turner and Jordan. That remains an area for further research and discussion.

One topic that *is* addressed by most of the authors in the book is the place of humans in nature. Surprisingly, even many of those who reject ecological construction as a new paradigm acknowledge the value of including humans in nature. Seeing humans as one among many species in the panoply of evolution is an integral part of the Turner/Jordan paradigm. Their insistence upon human responsibility to restore and propagate the rest of nature proceeds from the premise that human action is an inextricable part of nature, making strict preservation of wilderness areas from human influence impossible. Dwight Baldwin, Kim Medley, and John Wierwille, for example, cite widely varied situations in which they champion the adoption of new and constructive human interaction with highly degraded landscapes. Even Stanley Kane, who holds an essentially preservationist position, finds the idea of human membership in a community of species far preferable to exclusion. For Kane the model of human dominion over nature that we have inherited is based upon a presumed opposition of humanity and nature. Nonetheless, he warns that there are several ways to conceptualize human membership in nature, and urges that we not think of humans as the most privileged members of the community. However that issue may be resolved, the book does suggest that the Darwinian lesson, understanding humanity as an integral part of nature, is finally being learned.

Another issue alluded to by many authors but not resolved in this volume is economic and political. Humanists and scientists alike remark upon the fact that we will have to prioritize among preserving, restoring, and constructing ecological systems. As we do not have unlimited time or material resources, political decisions will inevitably be made, and they will probably be made in some degree of harmony with the popular mind described by Dora Lodwick and with some regard to the technical abilities and ideological proclivities of the community of environmental scientists discussed by Gary Barrett. But being human decisions, they will also be made in contexts where all the other needs of human society must be weighed and satisfied, from the most immediate needs of the population for shelter and sustenance to the perceived long-range needs for economic development, military defense, and so on. It is worth noting that a new ecological paradigm will by definition be more comprehensive and thus more difficult to put into practice than a new paradigm in physics or biology. Whatever new paradigm emerges, it will almost certainly involve a reordering of all our priorities. Perhaps it will be such a comprehensive change of mind and worldview that it will be compared to the great religious awakenings of the past. In that case, myth and ritual will certainly come along with political and economic decisions.

Much remains unexplored by this volume. Most obviously it is not a full set of case studies of different types of ecosystems and how they may be restored. Nor does it attempt to cover all the areas of the globe. It is not a manual for restoring or inventing landscapes. It does, however, offer a wide range of perspectives on one proposed ecological paradigm: that of ecological restoration and construction proposed by William R. Jordan III and Frederick Turner.

The value of a grand innovation or new paradigm is not often apparent to those who first encounter it. Indeed, it is a commonplace of the history of science and of intellectual history generally that most large-scale innovations are resisted at first, especially innovations that entail a thorough conceptual reorientation. They are most heartily resisted by those trained and working in fields where the new paradigm directly applies. This might give pause to those who are inclined to dismiss the positions of Turner and Jordan. The inability of the larger public to appreciate the alleged exhaustion of the preservationist paradigm does not invalidate their ideas, although it may delay their implementation: thinkers and ideas "ahead of their time" are equally well known phenomena. Yet these caveats do not validate the constructionist position either. How are we to know if ecological construction is a path-breaking idea whose time is only dawning, or a sidetrack that might lead us irretrievably astray? Only a future of much research, discussion, and practical application will tell us with any certainty.

The greatest incentive to continued research in restoration and further discussion of the constructionist paradigm lies in the immensity of the consequences. Everything is at stake. The current ecological crisis has provoked a conscious search for new patterns of human relations with nature that has no parallel in the past, at least not since the invention of agriculture. Ecological restoration, construction, and invention as advocated by Jordan and Turner are certainly not the last words that will be spoken in the welter of proposals elicited by this crisis. But their positions have special virtues. Not least, they constitute a clearly delineated and systematic alternative to conservationist thinking as we have known it. They point up the alternatives more dramatically than anyone has yet done. With their gamut of ambitions, from restored tall-grass prairies to terraforming Mars, Jordan and Turner have given us a glimpse of how differently the world may look from the perspective of a new ecological paradigm. And *some* new paradigm is surely in store for us. Therefore, as a research program for scholars in many disciplines, and as food for thought for thinking people generally, the ideas of Jordan and Turner as well as the criticisms published here are an essential experiment in thought. They cannot be lightly dismissed.

Contributors

A. Dwight Baldwin, Jr., earned his B.A. from Baldwin College in biology (1961), his M.S. from the University of Kansas in geology (1963), and his Ph.D. from Stanford University (1967). He currently holds the position of professor in the Department of Geology at Miami University, where he teaches courses in hydrogeology and water resources. In 1969 he helped establish the Institute of Environmental Sciences at Miami University. His publications include *An Evaluation of River Restoration Techniques in Northwestern Ohio*, a coauthored report to the Army Corps of Engineers.

Gary W. Barrett is a distinguished professor of Ecology at Miami University, where he founded the Institute of Environmental Science and Ecology Research Center. He is the author of two books and over one hundred publications in major scientific journals. He has been Ecology Program Director with the National Science Foundation, president of the United States Section of the International Association for Landscape Ecology, a member of the Board of Directors of the American Institute of Biological Sciences, chairman of the Applied Ecology Section of the Ecological Society of America, and a fellow of the American Association for the Advancement of Science.

Ann Cline holds degrees of Bachelor of Architecture from the University of Washington and Master of Architecture from the University of California at Berkeley. During the Carter administration, she directed a pilot community energy planning program for the U.S. Department of Energy. She currently teaches and practices architecture at Miami University. Her special interest is the translation and design of the teahouse, and she currently holds research awards from the Graham Foundation for Advanced Studies in Fine Arts and the Ohio Arts Council. Her most recent teahouse, designed as a miniature art gallery, will be onsite at two public exhibitions in 1993. Recent publications and conference presentations explore the connection between the rituals of the gallery and the teahouse, and their relationship to art theory and ecological strategy.

Judith de Luce is professor of classics, affiliate in women's studies, and fellow in the Scripps Gerontology Center at Miami University. She earned an A.B. in Latin at Colby College in 1968, and completed her Ph.D. in classics at the University of Wisconsin-Madison. Her research has focused on opportunities to define what it means to be human and to reflect on what behavior is more characteristic of humans than other animals. She is coeditor, with Hugh T. Wilder, of *Language in Primates: Perspectives and Implications*, and coeditor, with Thomas Falkner, of *Old Age in Greek and Latin Literature*.

David Gorchov received his Ph.D. in biological sciences from the University of Michigan and currently is an assistant professor of botany at Miami University. He has studied the ecological interactions between plants and animals that disperse their seeds in Michigan, New Jersey, and Costa Rica. In collaboration with Peruvian scientists, he leads an investigation of the regeneration of rain forest after strip clear-cutting in the Peruvian Amazon.

William R. Jordan III has been outreach officer at the University of Wisconsin-Madison Arboretum since 1977. He is the founding editor of the journal *Restoration & Management Notes*, and a founding member of the Board of Directors and Supervisor of Administration for the Society for Ecological Restoration. He is coeditor, with Michael E. Gilpin and John D. Aber, of *Restoration Ecology: A Synthetic Approach to Ecological Research*.

G. Stanley Kane is professor of philosophy at Miami University. He studied philosophy and religion at Harvard University, receiving his Ph.D. degree in 1968. He has held teaching appointments at the University of Wisconsin–Stevens Point and Illinois State University. He is author of *Anselm's Doctrine of Freedom and the Will* and of articles on the history of philosophy and the philosophy of religion.

Jack Temple Kirby is W. E. Smith Professor of History at Miami University, specializing in rural life, agriculture, and environmental history. He is the author of *Rural Worlds Lost: The American South, 1920-1960* and editor of the Studies in Rural Culture series of the University of North Carolina Press.

Dora G. Lodwick is associate professor of sociology at Miami University. Her areas of specialization include environmental sociology, social impact assessment, demography, and research methods. Her latest publications include *Viewing the World Ecologically*, coauthored with Marvin E. Olsen and Riley E. Dunlap.

Orie L. Loucks is professor of zoology at Miami University. He has worked on understanding the mechanisms and stability of ecosystems, particularly of forests, prairies, and wetlands, with emphasis on the contributions of individual species to the patterns of disturbance and recovery so characteristic of stability.

Kimberly E. Medley is an assistant professor in the Department of Geography and an affiliate in the Department of Botany, Miami University. Her academic background is multidisciplinary with a B.S. in conservation, M.A. in geography, and a Ph.D. in botany, and has been enhanced by forest-related work experiences in the National Park and Forest Service, The Nature Conservancy, and research projects in Michigan, Malaysia, Kenya, Ohio, and the New York City metropolitan area. Her research and teaching interests focus on plant geography, environmental and human influences on forest ecology and the conservation of natural resources.

Constance Pierce is a professor of English at Miami University. Her writing has appeared in *Michigan Quarterly Review* and *SubStance*, and she has published a story collection, *When Things Get Back To Normal*. Her work has been supported by fellowships from the National Endowment for the Humanities and the National Endowment for the Arts.

Carl Pletsch is associate professor of history at Miami University, where he teaches European intellectual history. He received his Ph.D. from the University of Chicago in 1977 and taught at the University of North Carolina at Chapel Hill from 1978 to 1985. He has been a member of the Institute for Advanced Study, Princeton, and Mellon fellow at the University of Pittsburgh. He is the author of *Young Nietzsche, Becoming a Genius* and essays on other figures in modern intellectual history.

Ellen Price is an artist and assistant professor of art in printmaking at Miami University. Her prints and drawings have been displayed in numerous invitational and competitive exhibitions, including the International Biennial Print Exhibition in Taipei, Republic of China, and the Philadelphia Print Club's 64th Annual Exhibition. Born in New York City, she studied life drawing at the Art Students League, and attended Hunter College and Brooklyn College, where she graduated in 1980 with a B.F.A. She received her M.F.A. in printmaking in 1986 from Indiana University in Bloomington. Her chief area of expertise is intaglio printmaking, specifically etching and drypoint.

Frederick Turner is Founders Professor of Arts and Humanities at the University of Texas at Dallas. He was educated at Oxford University and has taught at the University of California at Santa Barbara and at Kenyon College, where he was editor of the *Kenyon Review*. He is the author of over a dozen books of poetry, criticism, and fiction, including *Natural Classicism: Essays on Literature and Science; Genesis: An Epic Poem; Rebirth of Value: Meditations on Beauty, Ecology, Religion and Education*; and *Beauty: The Value of Values*. He is a regular contributor to *Harper's Magazine* and has been interviewed on two PBS Smithsonian World documentaries.

John E. Wierwille received his B.Ph. from Miami University in 1992. The focus of his studies was American environmental history. Prior to completing his undergraduate degree, he worked for three years with the international environmental organization Greenpeace as a community organizer and activist. His primary interests are advocacy and education.

He is currently a first-year student at the Emory University School of Law.

Gene E. Willeke, a civil engineer with interests in environmental, socioeconomic, and hydrologic aspects of public-works planning, has directed the Institute of Environmental Sciences at Miami University since 1977. He received the M.S. and Ph.D. degrees in civil engineering from Stanford University and has taught at Stanford, Georgia Institute of Technology, and Miami University. He has been a consultant to several federal and state agencies, notably the U.S. Environmental Protection Agency and the U.S. Army Corps of Engineers. He currently directs the preparation of a drought atlas of the United States for the Corps of Engineers.

Index

Compiled by Eileen Quam and Theresa Wolner